James Ballou

Arming for the Apocalypse

I0104337

Arming for the Apocalypse:
Assembling Your Survival Arsenal . . . While You Still Can
by James Ballou

ISBN: 978-1-943544-07-3

PrepperPress

Published by Prepper Press
Post-apocalyptic Fiction & Survival Nonfiction

www.PrepperPress.com

Contents

Warning

The information in this book is based on the experience, research, and beliefs of the author and cannot be duplicated exactly by readers. Neither the author nor the publisher assume any responsibility for the use or misuse of information contained herein. The contents of this book are intended *for academic study only.*

Acknowledgments

This book would not have been possible without the help of friends and the people who generously lent their time and support to the cause, and in some instances provided hard-to-find weapons from their own collections to be studied and photographed.

I owe a huge debt to two individuals—Don McLean and Jon Ford—whose combined wisdom steered me toward a much more intriguing approach with this project than that which I had initially started.

Special thanks go to the following individuals: Knut Rogers; Wyatt Rogers; Jerry Diemer; Matt Kelso; Mark Anderson; Robert Lenhart; John Tanner; Dion Unruh; Daniel Farmer; William Frank Lawson; Jacob Stewart; my wife, Alicia; my brother, Marty; and my son, Eugene Ballou, all of whom helped me obtain the photo images I needed, and especially my dad, Gene Ballou, who helped me tremendously with the book's gun projects, among so many other things.

I also owe my gratitude to the firearms manufacturers who graciously provided me with images of their products when I asked for them, including Thompson Center Arms, Heizer Defense, and Pietta of Italy.

Finally, I wish to say thank you to my publisher, Paladin Press, for turning this concept into a book.

Thanks all!

Introduction

I can easily envision the unpleasant scenario where civilization falls apart for any of a variety of reasons, perhaps even on a global scale, where anarchy could rule the streets of almost every major city following an apocalyptic event. In the chaotic social atmosphere that would ensue, self-preservation and the security of one's family would be high on just about everyone's list of priorities. Regardless of what some people may think about guns or other kinds of weapons right now, I am confident that the majority of citizens would prefer—even be eager—to arm themselves under such circumstances. People would suddenly become very conscious of food acquisition when there are no grocery stores, and of self-defense when there are no police.

I realize there have been books and magazine articles published in past decades devoted to the general topic of "survival guns." The question that begs asking, then, is just what new and compelling information could possibly be added to this discussion now? My short answer is *plenty!*

This book takes a somewhat different view of the topic than what readers will find in other books. Here, we will ponder such things as how we might obtain surprisingly useful, life-saving service from often-overlooked older or low-value firearms; make various modifications that could make our firearms more suitable for postapocalypse survival; turn our chosen survival firearms into actual "survival kits"; and even consider alternatives to conventional firearms, to list just a few examples. Additionally, we've seen quite a variety of interesting gun-related developments within just the last 15 or 20 years, so at the very least, an update of the topic is in order.

Admittedly, any debate we could wrangle over gun-related issues in this context has been done already and at length, and any opinions we could formulate concerning some of the more established conventional firearms have long since been adopted by some and dismissed by others. But, of course, that would be the case with just about any popular topic, and I am confident that readers of preparedness books overwhelmingly crave literature pertaining to firearms (as I do). Given how truly important weapons (and weapon choices) could become to survivors who may have to hunt for meat or defend themselves against numerous postapocalypse threats, the more we must force ourselves to consider all the pertinent factors and make exhaustive comparisons now, before the flag goes up. Because when the world around us starts to crumble, it won't be time to select firearms; it will be time to lock and load!

Oh, what a can of worms we will be opening up in the pages of this book!

CHAPTER 1

How to Begin Our Selection Process of the Ultimate Weapons for the Apocalypse

I n some respects, the prospect of deciding upon one particular weapon, or ideally a small selection of weapons, for extreme, desperate times might be easier for the novice than for many long-time firearm enthusiasts. For one thing, the choices ultimately made by someone learning from this book will be influenced in large part by this whole apocalyptic vision we will be pondering. Our choices have to be based entirely within the context of desperate survival rather than on the nostalgic or sentimental reasons some of us who have owned and used firearms for many years might be inclined to apply in our decision making, either knowingly or unknowingly. This methodology of putting functional utility ahead of other priorities will be underscored throughout this book.

An example of a sentimental priority governing one's decision about weapon choice would be the individual who has always hunted deer with the same old octagon-barreled lever-action rifle that his dad always hunted with, and maybe even his dad's dad, for many years. Whenever he takes that particular rifle to the woods, he relives the memorable hunting trips with his father and grandfather. As someone who tends to get sentimental about this very sort of thing, I can personally relate.

I also have a special fondness for those early lever-action rifles like the one in that example, having been fascinated with just about everything associated with the American West since the days of my youth. The family heirloom rifle may indeed work very well for that individual and it might even be the perfect all-

A small assortment of weapons that could become useful after the apocalypse.

around weapon for the apocalypse—but for *him*. For our discussion, we will at least endeavor to consider the topic more objectively, and from the right angles.

For example, when considering survival firearms, there is one occasionally neglected factor that is every

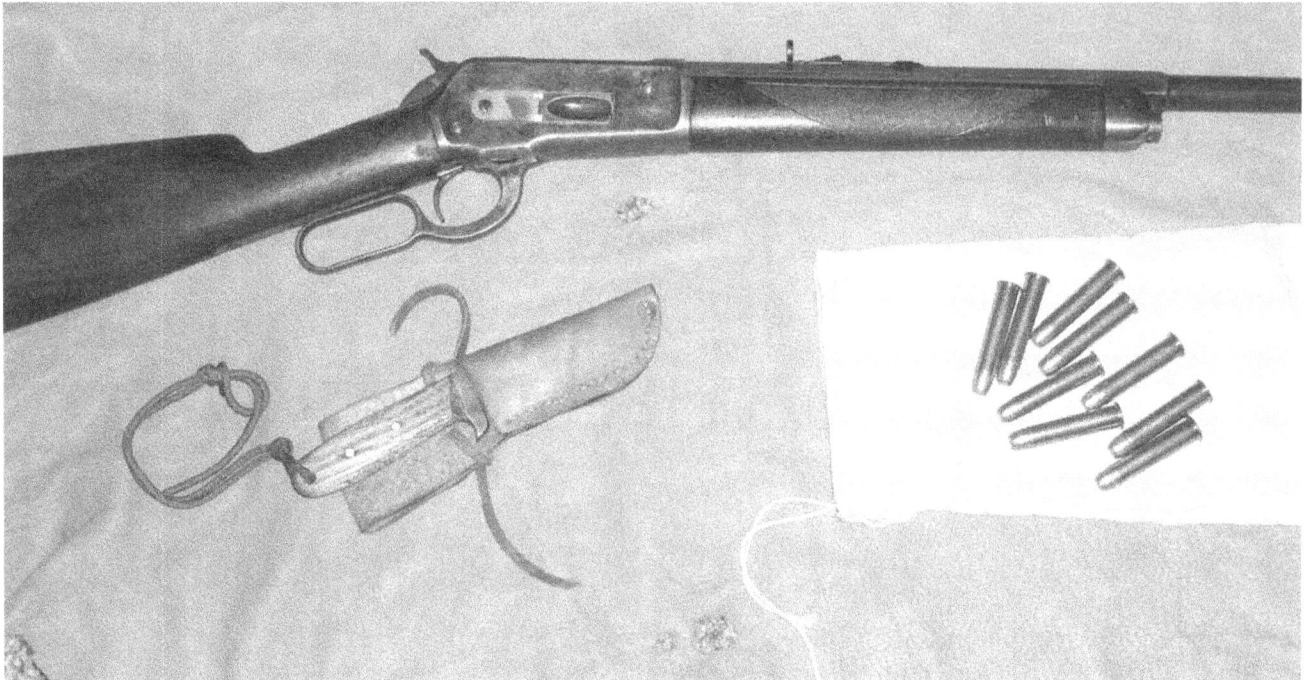

An old rifle such as this Winchester Model 1886 in .45-70 might be perfectly suitable as a postapocalypse survival gun, especially for the guy who has hunted with it all his life.

bit as important as the specific *type* of weapon ultimately selected. It is that a shooter who shoots the same gun all the time will have the advantage of familiarity with his weapon. This can be a huge advantage in a crisis, because familiarity bolsters confidence.

For this simple reason, our nostalgic deer hunter is probably well advised to stick with that old rifle he's hunted with all his life, because he has a solid understanding of its capabilities and limitations. The most ideal theoretical weapon for the apocalypse that might ever be invented won't do much good at all for the individual who hasn't spent the time to learn how to use it effectively. By contrast, the guy who has been shooting the same gun for years will have a pretty clear sense about its capabilities, will have become accustomed to all of the gun's idiosyncrasies, will have developed a feel for its trigger, will know what kind of sight picture to use at various distances, and so on. There really is no substitute for familiarity with the equipment one must depend on in an emergency.

To develop this point further, another benefit of shooting the same gun all the time is the element of consistency. The shooter becomes accustomed to the same set of sights; the same trigger pull; the same recoil; the same balance, feel, and weight characteristics of the weapon; and so on. His shooting results should therefore be more consistent than if he were to

fire a different weapon every time he went shooting or frequently switch between guns at the range.

Years ago when my dad and I would routinely venture out past the edge of town during summer weekends and spend whole afternoons plinking, we would more often than not bring several guns with us, and quite often those would be newly acquired firearms. We both did our share of gun trading and collecting in those days, and we naturally were eager to test the performance of any newly acquired models. But it would be a learning curve each time, trying to get them sighted in and developing a feel for each individual weapon. Life is much simpler—and more consistent—for the one-gun man.

Weapons for a postapocalypse world won't usually be the prettiest or showiest. They won't have to win any competitive sport shooting matches, they don't have to be the biggest or the most powerful, and they won't have to impress your friends. Their sole purpose will be to help you survive under desperate conditions.

Some of the common older guns might not be as exciting as a lot of the newer choices, but many of them were reliable workhorses in their day, and those weapons could still serve us quite well in desperate times. Personally, I think the World War I-era British Lee Enfield, No. 1 Mk. III SMLE (Short, Magazine, Lee Enfield) is one of the homeliest rifles I have ever seen, but I also know that it was a solid, durable, and

Any Russian or Chinese-made SKS semiautomatic rifle is considered an ugly duckling by many, but the gun pictured here is reasonably accurate at 100 yards, and it never fails to fire when needed.

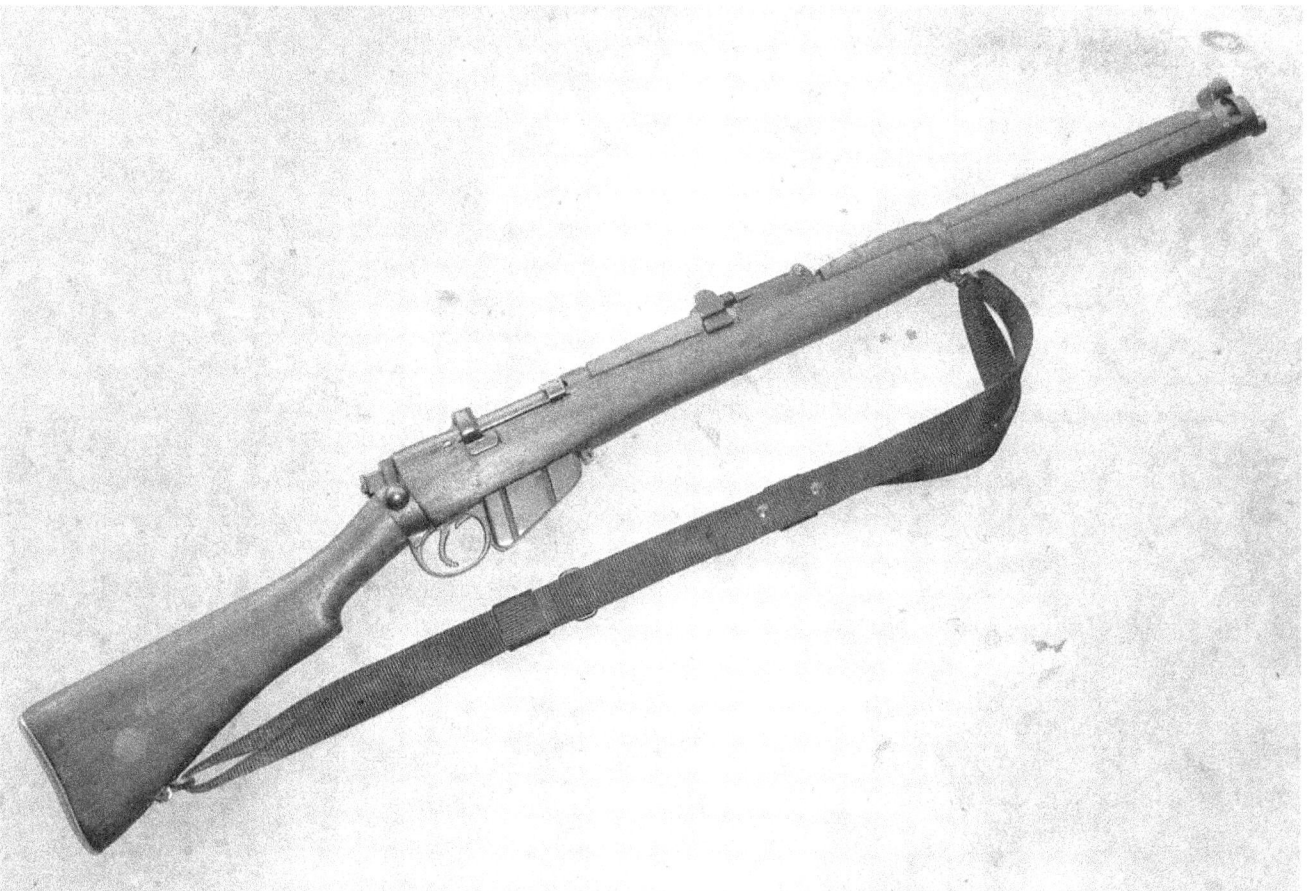

The British Lee Enfield SMLE bolt-action rifle used during World War I, caliber .303.

very reliable weapon that served the Crown quite well in its day, and the Brits who used it against the Germans often demonstrated a remarkable degree of proficiency with it. These are the same characteristics you want in your survival guns.

The process of making decisions based solely on the information in this book may actually be simpler for some readers than the same process supported by a more comprehensive body of information. One of my goals is to simplify the topic for readers as much as possible and to help make the decision-making process *easier*. We are going to endeavor to reason our way through this whole process in these pages.

This search for ideal general-purpose survival weapons has engaged my thoughts for nearly 40 years, and I sure wish I had a dollar for each time I've formed my opinions about one aspect or another and then later changed my mind. There are even plenty of firearms enthusiasts and survival experts who will insist that the perfect "all around" survival weapon does not exist.

One of the major dilemmas involved here lies in the infinitely wide range of potential survival scenarios in so many different parts of the world that would constitute vastly different requirements.

For instance, a trapper living in a log cabin way back in the bush of Alaska would likely want at least one high-powered, heavy-caliber rifle, perhaps something like a .338 Winchester Magnum, a .340 or other heavy Weatherby Magnum, a .45-70 or .450 Marlin Guide Gun, or maybe even a .375 H&H with which to kill bears or moose. An urban survivor, on the other hand, might not need anything quite as powerful but instead may have a need for a fast-handling, close- to medium-range handgun, shotgun, or carbine capable of providing pest control inside an attic or abandoned warehouse, as well as self-defense against other armed people or feral dogs, inside or between buildings or across rooftops.

Someone getting along in a rural farming community may only need a shotgun or small-caliber rimfire rifle to harvest small critters like pheasants, rabbits, mink, or squirrels. The survivor who finds himself in a wide-open environment like the southwestern deserts or the central plains will likely want the flattest-shooting long-range scoped rifle he can acquire, while a survivor in a dense forest or jungle may wish for a short-barreled shotgun loaded with heavy buckshot. So we can see here that the unique scenarios and potential requirements for our survival firearms can be incredibly diverse, making it nearly impossible to strictly define our weapon's intended purpose.

Shooting-related products available today are also incredibly diverse, as we will observe repeatedly throughout this book, and as we strive to arm ourselves wisely for the coming apocalypse, we will be considering seemingly endless trade-offs—always contemplating sacrificing one benefit or another on the one hand in order to keep what we may consider an "essential" on the other.

GENERAL-PURPOSE VS. SPECIALTY WEAPONS

We've all heard the popular advice about using the right tool for the job. It comes from hard-earned experience and is certainly applicable in a modern industrial society. A screwdriver, for example, will turn a screw much more efficiently than will the point of your pocketknife. And it is generally easier to turn a nut on a bolt using the corresponding wrench or socket size than trying to accomplish the same task with pliers. It behooves us to grab the right tools for whatever we need to accomplish.

However, when the supply infrastructure of tools and equipment collapses, as we might expect to occur following an apocalyptic event, people will be forced to operate with whatever tools they have available, or in some instances be forced to improvise their own tools.

This will also be true with weapons. For example, a white-tail deer may present itself at two or three hundred yards, arguably outside of normal handgun range, but the apocalypse survivor may find himself armed with only a handgun (if he's even that lucky), and he may have to try to make it work for the task at hand. Likewise, he may find himself having to defend his home and family with an unwieldy full-sized hunting shotgun intended for shooting ducks on a pond rather than home defense. The bird gun may be awkward to move with through doorways, or swing the barrel quickly onto the target in close quarters, and those light birdshot loads aren't exactly ideal combat loads, but in that situation they must be used to get the job done.

So those two very different examples underscore our need for versatility in our weapons. By planning ahead with this kind of application crossover capability in mind, we might have a better chance of selecting the most versatile general-purpose weapons for the apocalypse.

In my view, there is no better time than the present

Selecting just a few practical arms for apocalypse survival out of any modest gun collection could be quite a challenge without methodical and thorough consideration before the desperate times arrive.

to consider this topic in depth. We must formulate at least *some* of our own ideas now, in this hypothetical context, rather than later, when society is crumbling and critical decisions will have to be made quickly and while available time or access to equipment could be severely limited.

Imagine just for a moment that you are one of the many firearms enthusiasts in our society who has accumulated dozens of guns over the years (as many of the readers of this book will have done no doubt), and the social conditions of civilization have finally overheated to the point where you are forced to abandon your metropolitan home in a desperate hurry. Let's just say for the sake of adventure that it happens to be in the middle of the night, when mobs of anarchists are burning down the houses in your neighborhood and you need to bug out with your family for a safer place away from immediate danger.

You suddenly discover that your load-carrying capacity will accommodate maybe five or six weapons at most, with a fairly limited supply of ammunition for each, due to the need to also bring food, clothing, water, sleeping bags, tools, tent, and other essential supplies that fill up your SUV. Regardless of how fond you might be of your dozens of great guns, situational demands and logistics are what they are, and now you find yourself forced to make some tough decisions.

So there you are, frantically packing up everything you believe you will need for the long haul while houses on your block are erupting in flames, and you have to ask yourself that question: which guns do you take and which do you leave behind?

It was imagining this hypothetical scenario that inspired me to write this book. When push finally comes to shove, which single firearm or which few weapons from a broader selection of potentially very useful

My dad's hypothetical three-gun selection for the apocalypse: his custom-built .25-06 scoped sporting rifle that he has hunted with for many years, the LC Smith 12-gauge side-by-side shotgun his father gave to him for his sixteenth birthday, and his favorite revolver that he has owned for nearly 20 years, a Smith & Wesson Model 57 .41 Magnum.

weapons should someone who is bugging out in a hurry grab *first*?

I don't believe this is a puzzle we could properly solve simply by inputting all of the relevant data into a computer and having the machine calculate the best answer. Our matrix of variables is a bit more complex than that!

In theory, the ideal weapon would have to be capable of protecting you from such dangerous wild beasts as bears, cougars, wolves, starving feral dogs, the occasionally aggressive bull moose in the northern woods or wild boars or alligators down south, a charging plains buffalo or an angry rogue steer, dangerous exotic animals that may have escaped from zoos (it has happened before, and it could certainly happen after a societal breakdown), or any other dangerous creature in the animal kingdom, as well as from any hostile humans trying to do you harm. So this requisite more or less forces us to dismiss small-caliber, low-powered firearms that might otherwise be ideal for hunting

small game for food, at least in our search for those first-grab, general-purpose weapons.

The ideal weapon should be at least as effective as any weapons possibly used against you. It should be reliable under the harshest of conditions and in all weather extremes. It should be equally capable as a hunting tool as it is a defensive tool when needed, to help you collect meat, suitable for taking everything from small animals and birds to the largest animals encountered in your region. And ideally, whatever weapon we choose should be portable, sustainable, easy to use effectively, and, to reiterate this critically important trait, *reliable*.

As we can see, selecting any single weapon or even a small assortment of weapons to fulfill all of these requirements is indeed a tall order. But of course, we have to start somewhere if we're going to talk about it at all.

One common approach is to select three primary firearms—one from each general category—so that we

The author's idea of a reasonable selection for the apocalypse from the three main categories of firearms: a semiautomatic rifle in .308 caliber, a pump-action shotgun in 12-gauge, and a GLOCK 10mm pistol.

start with one handgun mainly for close-quarters self-defense, one rifle mainly for hunting and taking longer shots accurately, and one shotgun for more of a multi-purpose close-range tool.

This logical and seemingly simplistic approach does address a relatively broad spectrum of potential circumstances. Where it gets involved is when we begin to ask ourselves specifically *which* handgun we would choose, *which* rifle best fits its own niche, and *which* shotgun we should decide upon to round out our three-gun arsenal for the apocalypse. As we've already acknowledged, the various designs, brands, sizes, action types, sighting systems, and caliber options are vast.

It may be beneficial to create a list of our chosen apocalypse weapons in some order of priority and post it someplace where we can find it in a hurry should the need arise, but ideally someplace not easily accessible to others. On the inside of our gun safe door might be a logical place for such a list (for those who keep weapons in a gun safe), because the list won't be accessible to others while it is in there, and we'll have to get into the safe to access our weapons anyway.

One of the reasons why I think it makes sense to create a list, and to update it periodically, is because doing so will force us to put a lot of thought into the issue, and it will help us organize our overall preparedness plan. Additionally, it is very easy to forget details over time, and it sure would be a shame to invest a lot of time and research into all of this, weighing all of the pertinent variables and finally formulating a basic, logical plan, only to forget three to five years down the road much of what we had decided, when those decisions suddenly become critical. (Obviously, this would be more of a concern for a gun collector or someone with a dozen or more weapons than it would be for the hunter who owns only a few sporting arms.)

Another possibility is to develop a chart that is not merely a list of specific weapons that we believe would be the most suitable for the apocalypse but

A fairly small supply of assorted ammunition can weigh several hundred pounds, so along with everything else, we have to concern ourselves with the portability and storage of ammo and other supplies that support our weapons.

perhaps a list of weapon types arranged categorically. This is more of a generic approach that I believe could be very helpful to us later, at least in helping us avoid forgetting some of the most important factors when electing doomsday guns. The main advantage to this kind of general list (over a list of specific individual weapons) is that it would not require as many updates for those who occasionally trade or sell their guns or periodically acquire additional pieces as time goes by.

We may also wish to create a chart that would serve as a guide to the advantages and disadvantages of all our different weapons or weapon types. If we are faced with having to make difficult decisions about weapon selection for the apocalypse at some point, a

kind of pros vs. cons chart might be helpful in making the best decisions under pressure.

There are, of course, other related issues besides just the weapons themselves that must be factored. For instance, ammunition requirements will be one of our major concerns, because ammunition of almost any kind in large quantity is heavy, often bulky, and can be very expensive, not to mention the whole issue of availability for resupply after the collapse of civilization.

So, let us begin weighing all of these many factors and formulating some semblance of priority out of all of it. Our goal is to choose firearms that we have confidence will work for us when the moment of need is at hand.

CHAPTER 2

Choosing the Apocalypse Handgun

I will start this sorting-out process with the handgun, simply because it is the one type of weapon (other than maybe a knife) that, due to its compact size, you could conceivably always have with you, on your person in a holster or otherwise within easy reach, regardless of whatever activities you might be engaged in.

It is appropriate here to carefully consider the fundamentals of some of the different handgun types for this discussion. That age-old revolver vs. semiautomatic debate keeps rearing its ugly head. And within both categories, revolver and automatic, we also find a vast array of calibers, styles, barrel lengths, cartridge capacities, sighting systems, composition materials, grip shapes and sizes, overall sizes and weights, and so on.

Once again, all of these variables can seem overwhelming to folks not familiar with firearms who haven't already developed their own opinions or biases. And without actually handling and test firing guns of each variation, any decisions must rely on the information available—which is hopefully based mostly on objective, unbiased data. So that is what we will attempt to explore here, although we cannot entirely escape personal opinions with a discussion of this nature, and I will certainly share some of my own in these pages.

Many shooters believe that, as a general rule, the revolver is inherently more reliable than the automatic. Now, we could debate the merits of this notion until the cows come home, and indeed it has been endlessly debated for eons, but what we might all ultimately agree on is that the reliability factor should be of

Some shooters prefer revolvers, while others prefer automatics. Which type of handgun will you choose for the apocalypse?

paramount importance to all of us preparing to arm ourselves for the worst of situations. So, regardless of whether we select an auto or a wheel gun, we want our choice to be as reliable as a firearm in its category possibly can be.

IF WE SELECT AN AUTOMATIC PISTOL

Naturally for some shooters, and certainly for some purposes, semiautomatic pistols will be more suitable than revolvers. And if we are to consider one

Four examples of auto pistols, clockwise from upper right: Springfield 1911 .45 ACP, Ruger LCP .380 Auto, Czech CZ-52 in caliber 7.62x25, and a 10mm GLOCK 20.

Feeding a loaded magazine into an auto pistol is normally faster and more convenient than re-loading a revolver's cylinder.

particular automatic or another as our main sidearm, we need to ask ourselves five basic questions . . .

Question #1: Is the pistol *easy enough to operate and shoot?* In other words, is it ergonomic for us, can our fingers reach the controls quickly and comfortably, and can we easily aim, fire, and reload the weapon? If the slide is difficult for a small person to grip and pull all the way back to cock the weapon for the first shot, for example, then another handgun design might be better for that person.

Just as with rifles and shotguns, some handgun designs simply fit certain shooters better than others and will be easier to use efficiently and effectively. You should be completely comfortable with the handling characteristics of your weapons, and the best way to determine which gun is best for you is to handle as many as possible at a gun shop or gun show and to test fire your favorites at the range. Many indoor shooting ranges rent a variety of popular handguns.

Question #2: Is the pistol *powerful enough* for the task for which it is intended? It would be awfully discouraging to fire a weapon in desperate self-defense and suddenly discover that its bullet was ineffective against the aggressor.

Let's dig deeper into this issue of "power," as the term can imply different meanings. Shooters talk about "knock-down" power, "stopping" power, "bullet energy," and "power factor," but we'll need some kind of basis for reference when comparing the capability of the various cartridges.

Some auto pistol cartridges are considered effective man-stoppers by researchers who study the science and statistics on this, and other cartridges are considerably less effective for this purpose. (The same will naturally be true for the whole lot of revolver cartridges as well.) And self-defense against hostile humans is, of course, just one particular application to consider.

On the more powerful end of the auto pistol scale, we have the .45 ACP (especially with the +P loads), the 10mm Auto, the .357 SIG, the .38 Super, both the .41 and .50 Action Express, several other cartridges that never became commercially popular like the .44 Auto Mag and the .45 Winchester Magnum, and perhaps even the .40 Smith & Wesson.

On the more anemic, or at least the lighter end of the scale we find the .25 ACP, .32 ACP, .380 Auto, .30 Luger and the similar but hotter 7.62x25 Tokarev, 9mm Makarov (9x18), and perhaps a few others. Somewhere in the middle we might place the recent experimental .45 GAP, the now almost obsolete .38 Automatic, and most of the 9mm loads on the market.

In my view, it would make sense to select a sidearm chambered for one of the mid-range to more powerful cartridges as our main handgun (my favorite being the 10mm, for reasons I will explain a bit later), and possibly one of the lighter calibers such as .32 ACP or .380 in a much more compact hideaway or backup *secondary* weapon (as we will explore at length in chapter 10), if your load-carrying capacity allows for it.

Now let's take a moment to consider just exactly what we're actually talking about when we say that one particular cartridge is "more powerful" than another, because as already observed, power can be measured in several ways.

The muzzle energy of a bullet, meaning the kinetic energy of the bullet as it exits a gun barrel, is commonly expressed in the United States as foot-pounds (ft-lbs.) of force.

Manufacturers' ballistics charts will typically provide, in addition to the bullet's velocity in feet per second (fps) and its trajectory at various ranges measured laterally in inches, the muzzle energy in ft-lbs. for the various cartridges in their catalogs. You can also find this ballistic data for nearly every cartridge and common load in gun magazines and hand-loading manuals. There are even websites with on-line calculators that can calculate the figures for you.

Even so, the formula for calculating bullet energy could be useful to know for a time when computers won't be available. It is as simple as multiplying the bullet weight in grains by the velocity in feet per second squared, and then dividing that by 450,436. (Note: This constant is based on the now generally accepted gravitational acceleration of 32.174 ft/s^2, but this will not be the same everywhere in the world. Readers may encounter other constants, such as the 450,240 shown in older hand-loading manuals that used a different gravitational acceleration, or now sometimes 450,400.)

We can easily check our method. Winchester's 2012 ammunition product guide, for example, shows a 10mm Auto load having a bullet weight of 175 grains, with muzzle velocity of 1,290 fps and muzzle energy of 647 ft-lbs. To use our formula to confirm the muzzle energy, we multiply 175 by 1,290 squared (or 175 x 1,664,100), and that gives us 291,217,500. If we then divide that by the 450,436, we end up with 646.52, rounded up to 647. And it works every time.

Ironically, a bullet that delivers comparatively high kinetic energy is not necessarily always a bullet with equally reliable stopping power. It depends largely upon the physical characteristics of the target body, may in some instances actually penetrate the target's body completely. In so doing, it may fail to impart all its kinetic energy directly to the target, as would be ideal for maximum stopping power, since it carries some of its impact energy to its point of rest somewhere beyond. We will talk more about what we mean by "sectional density" in a little while, but the

point here is that the most effective man-stoppers usually tend *not* to exit the body but instead transmit all of their impact energy to the target.

And bullet design is a key component here. For example, a hollow-pointed bullet tends to resist over-penetration much better than a full-metal-jacketed bullet, and it also tends naturally to do more damage, i.e., makes a bigger hole and wound channel inside the target body than will a bullet of the same diameter that doesn't expand as much.

However, there are certain situations in which the maximum possible penetration of the bullet will be desirable, such as when attempting to punch through a light barrier of some kind in order to hit the target on the other side, or when we want our bullet to reach the vital organs of a particularly large, tough, and possibly dangerous beast.

In those situations, we would want a more solid bullet design and, in addition to a bullet traveling at the highest possible velocity, ideally a bullet with a comparatively high sectional density, meaning the

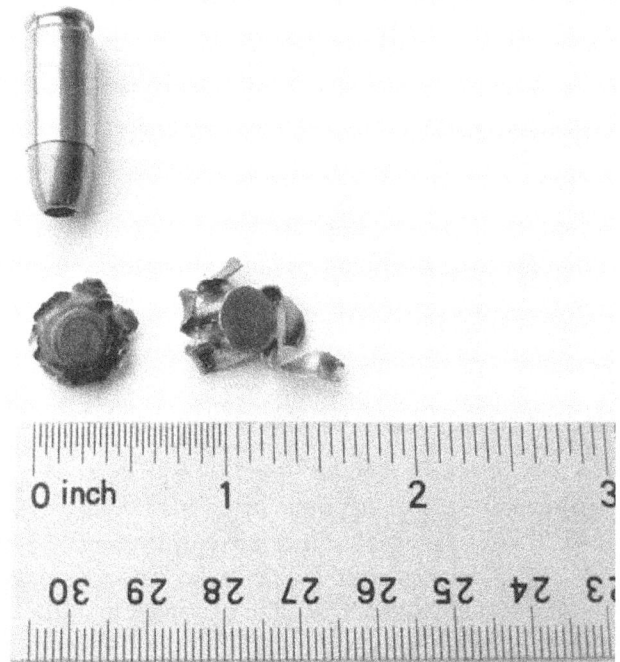

Shown here is a 115-grain 9mm jacketed hollow-point bullet that was fired into a pail of water that had a small stack of newspapers over its lid. Although the jacket at right has separated from the lead core at left, the lead portion mushroomed to .56-inch diameter. If similar expansion occurred in a human body, a considerable amount of tissue damage would be the result.

amount of mass with a given cross-sectional area, or SD = M/A, where M = mass, and A = area. As explained by Bob Beers in his article "Sectional Density for Beginners," as the frontal area (for our purposes, the bullet's caliber) increases, the weight behind it must also increase in order to achieve the same penetration.

Calculating the sectional density of any bullet is as easy as multiplying the diameter (caliber) of the bullet squared by 7,000 (the number of grains in a pound), and then dividing the bullet's weight in grains by that product, or perhaps more simply: the bullet's weight over 7,000, times bullet diameter squared. This is just nice to know whenever making comparisons of projectiles.

So we can see where the larger diameter projectiles also have to be longer in order to still possess the higher sectional density. And when we want to hit our target with both plenty of frontal area to achieve the highest knock-down power, while at the same time hoping for the deepest possible penetration for punching through car doors or maybe for reaching the vital organs of large animals, we will want to select a bullet with not only sufficient diameter and sectional density but that high-velocity component in there as well.

Revolver cartridges for handgun rounds, like the .41 and .44 Magnum, combine those three elements fairly well, whereas most of the auto pistol cartridges will be found lacking in at least one or two of these areas. The one exception that comes to mind is the 10mm Auto. The 10mm is .40 caliber (.400 inch), so the diameter of its bullet is slightly larger than the 9mm (.380 inch) but also noticeably smaller than a .45 ACP bullet (.452 inch).

With a 200-grain bullet, the 10mm has a higher sectional density (SD = 0.179) than that of a .45 with a 230-grain bullet (SD = 0.161). Additionally, muzzle velocity ranges for popular 10mm loads with the various bullet weights range from just over 1,000 fps typically to more than 1,300 fps, while only two out of the nine .45 ACP loads listed in Winchester's new catalog are shown with as much as 1,000 fps, and four of the nine are shown with muzzle velocities in the 800 range.

Additionally, the highest kinetic muzzle energy shown for the .45 ACP in that particular list is 432 ft-lbs. (with most of them hovering at 340 to 395), whereas a 200-grain 10mm bullet that exits the barrel at just 1,200 fps will deliver slightly over 639 ft-lbs. of kinetic muscle! Can you see now why the 10mm appeals to me as a likely candidate for an apocalypse handgun cartridge?

A .41 Magnum case case next to a 10mm case to show the dimensional similarities. The two cases are of nearly the exact same diameter.

Another, perhaps simpler method of establishing a basis for comparing the power of different loads is to calculate their "power factor." The power factor—a mass times velocity, momentum-based quantity (as opposed to an energy-based quantity)—is simply the product resulting from multiplying the feet-per-second bullet velocity, normally measured near the muzzle, by the bullet's weight in grains. For example, the power factor of a .45 ACP bullet weighing 230 grains that exits the pistol barrel at 850 fps would be 195,500 (230 x 850 = 195,500). The competitive pistol shooting world uses this method to categorize or rank different pistol cartridges for the different classes of competition.

We'll talk more about some of these cartridge comparisons and their performance in another chapter, but right now let's move on to question #3: Will our auto pistol selection *be among the most reliable*? Here is another subject that everybody seems to have an opinion about, but there are two factors I believe are worth weighing before making a determination. First, look at the *historical evidence*—has one model of pistol proven itself as a reliable tool more than others over time in enough real-world combat or equally harsh conditions? Second, how well has the pistol in question survived the *toughest man-made torture tests*?

I believe the legendary Model 1911-type automatic pistol invented by John Browning more than a century ago has answered the question about which pistol has proven itself better than any other ever produced. I can think of no other sidearm that has served

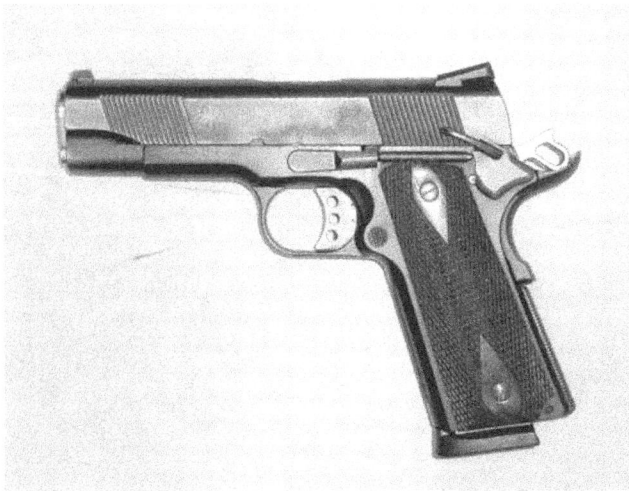

One recent version of the legendary 1911 pistol is Smith & Wesson's 1911 Sc. "Sc" is the element symbol for scandium, from which parts of this gun are made.

GLOCK 20, 10mm, at left, GLOCK 19, 9mm, at right. The magazines of both guns hold 15 rounds each.

a great nation's armed forces continuously for over 70 years, helped fight two world wars in addition to several other major conflicts, and has been copied as much or manufactured in greater numbers for as long a period as the basic Model 1911 automatic pistol, and the same gun appears to be as popular (maybe even more so) today as it ever was. Nearly every major handgun manufacturer in the world—including Colt, Remington, Smith & Wesson, Ruger, Iver Johnson, Taurus, Dan Wesson, High Standard, SIG Sauer, Springfield Armory, Kimber, Norinco, Para-USA, Thompson Auto-Ordnance, AMT, and Browning Arms Co.—is either making a 1911-type pistol at the present or has done so in the past.

While these points don't necessarily assure us that the 1911 is the most reliable pistol ever designed, its battle-proven history combined with its established reputation and enduring popularity certainly weigh heavily in its favor. You can't go wrong by exploring the dozens of quality 1911s as your postapocalypse handgun.

I believe that our second reliability factor—how well our pistol stands up to the toughest man-made torture tests—is best answered by the various automatic pistols made by the Austrian company GLOCK.

The GLOCK pistols were revolutionary when first introduced to the world in the early 1980s, because their frames, magazine bodies, and some of their small parts were constructed of polymer plastic. The guns are lighter in weight than comparable all-steel guns, and their frames are surprisingly durable and comparatively resilient to adverse weather and moisture. The

barrels and slides are made of steel, as are some of the internal parts (e.g., springs), with some of the steel parts wearing a protective coating on their surface. Overall, the GLOCK pistols have a reputation for being very rugged and reliable.

The torture tests originally conducted by GLOCK to demonstrate to the world that their pistols were durable included the famous ice test, dirt test, mud test, water submersion test, chemical degreaser test, and truck tire test. In the process, the plastic gun established its reputation as an undeniably durable product.

In the book *The Complete GLOCK Reference Guide* by PTOOMA Productions (a useful resource for any GLOCK owner), a GLOCK pistol (G23, .40 caliber) is subjected to a number of severe torture tests, including running it through 1,000 rounds rapid fire, shooting it with both extreme and no lubrication, putting it through a cement mixer, hammering it, submersing it in corrosive liquid, dragging it from a truck, shooting it with a .22-caliber rifle, and even firing it from a cannon!

Naturally, the tortured pistol didn't finish entirely unscathed, but it survived the testing remarkably well, still being able to cycle rounds even after enduring such unusual abuse. And there have been numerous other extreme conditions tests devised by shooters and organizations to drag GLOCK pistols to their limits, and time and again the pistol has impressed its testers.

One of the great advantages of the GLOCK family of weapons for our purposes is their inherent simplicity, and this also addresses that first question we asked

The GLOCK Model 19 disassembled into its five main component parts.

Pistol barrels with partially unsupported chambers: GLOCK 10mm at left and Beretta 92FS 9mm on the right.

about how easy the weapon is to operate. GLOCK pistols are all amazingly simple to load and fire, simple to disassemble and reassemble, and simple to clean and maintain. The gun features no external safety devices of the traditional configuration or complicated gadgets of any kind to fumble with during a stressful situation, when the weapon might be called upon in a hurry to save your bacon. For me, this is a huge selling point.

Of course, we might say that nothing revolutionary ever really comes without a fault, and not *every* pistol shooter is a huge fan of the "plastic gun."

For instance, hand loads are not recommended for use in a GLOCK pistol as it comes from the factory, for several reasons. Unlike most handgun barrels that have conventional cut-type rifling, the GLOCK barrel has polygonal-type rifling that tends to allow lead deposits to accumulate whenever firing a quantity of soft lead bullets. This buildup can cause increases in chamber pressures.

The GLOCK barrels also feature partially unsupported chambers in the feed ramp area, and especially hot hand loads—or reloaded cases that were previously fired in a partially unsupported chamber and thus excessively stressed—have experienced rupturing and blowing downward in that area (known as a "kaboom"), causing damage to the pistol and presenting a potential health hazard to the shooter. This actually happened to someone I know. He blew up his GLOCK 21 .45 Auto firing a hand load. Fortunately, he was not seriously injured.

Equally fortunate, the issue is easily solved by replacing the factory barrel with any of several

aftermarket barrels made for GLOCK pistols, having fully supported chambers and conventional rifling. This is important, because in a prolonged apocalypse, hand loads may become the only source of ammunition after factory and cached stocks run dry.

I should also point out that the partially unsupported chamber is not unique to GLOCKs. Since a beveled feed ramp tends to aid in the operation of an auto pistol, this feature is relatively common with automatics. The barrel of my Beretta 9mm also has a surprisingly deep feed ramp that results in a partially unsupported chamber, as is clearly visible in the above photo.

In the two separate 1,000-round rapid-fire torture tests that I am aware of—that first one I read about in *The Complete GLOCK Reference Guide* and the other I saw in a YouTube video (2008 Tactical Response Alumni Weekend Event)—the GLOCK's plastic guide rods melted in both instances after 900+ rounds had been fired. In the online test, the pistol, a 9mm GLOCK 19, continued to operate with some degree of reliability for the remainder of the test to eat up the last of the 1,000 rounds, even after the melted guide rod had fallen completely out of the gun!

The plastic guide rod seems to be a weak link in the GLOCK pistols. The factory guide rod in my own GLOCK 19 actually separated recently near the rimmed end that catches the spring during reassembly. I will most likely replace it with a steel rod instead of another GLOCK factory one, although I assume that this type of failure under normal circumstances is probably extremely rare. Aftermarket guide rods made of steel are available for any GLOCK owners wanting

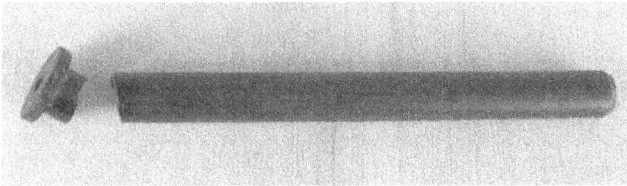

Broken factory plastic guide rod from a GLOCK 19 pistol.

The factory slide-stop lever is adequate on this GLOCK 19, but it can be difficult for the thumb to reach on a larger-frame GLOCK.

This aftermarket extended slide-stop lever makes the weapon much easier to operate with one hand.

a more durable part for peace of mind, or for anyone expecting to shoot a thousand rounds through a GLOCK in less than 14 minutes.

There is one small modification that I do recommend with any of the large-frame GLOCK pistols (mainly GLOCK 20s and 21s). The slide stops, or

The Beretta 92FS 9mm pistol.

slide-release levers on stock GLOCK pistols, are hardly more than just a flat sheet-metal tab exposed on the left side of the frame. This tiny factory lever is reasonably accessible to the thumb of most adult male shooters on the smaller GLOCK models, but my right-hand thumb—and I have fairly large hands—couldn't get an easy purchase on the lever of my GLOCK 20. I installed an aftermarket extended slide release to remedy this, and the pistol became incredibly easy to manipulate one-handed, without creating any sort of snag-prone protrusions on the gun.

As of this writing, extended slide-release levers for GLOCKs are available from Lone Wolf Distributors (www.lonewolfdist.com; search under "Custom Gun Works"), Aro-Tek Ltd. (www.arotek.com), and the Vickers version from NET TAC LLC (www.nettac.com).

Worth noting here are the recently introduced Generation 4 GLOCK pistols that feature interchangeable back strap options (which allow the guns to adapt to different shooters' hand sizes), enhanced texturing on the grip and slide, larger magazine-release buttons, and captive double-recoil springs that apparently soften recoil. These changes have been receiving a lot of praise from shooters.

When it comes to durability and performance tests, we should also take into consideration the testing conducted by the U.S. government in the 1980s that ultimately led to the selection of the Italian-made Beretta M9 (also the 92 series in the commercial market, with the Beretta 92FS having only subtle differences from the M9) that replaced the 1911A1 pistol as the standard-issue military sidearm more than 20 years ago

The SIG P229 shown here is the company's compact version of their popular P226.

Shooters should have no problem keeping their shots on paper within pistol range using a Beretta 9mm.

(effectively in the 1980s, but not formally until January of 1992). As explained on Wikipedia, "the Beretta 92F pistol survived exposure to temperatures from −40°F to 140°F, being soaked in salt water, being dropped repeatedly on concrete, and being buried in sand, mud, and snow."

Additionally, testing proved that the pistol could sustain 35,000 rounds before failure. In the U.S. Army's tests (including the 1988 XM10 competition), the Beretta beat out competing pistols from Smith & Wesson, SIG Sauer, Heckler & Koch, Walther, Steyr, and Fabrique Nationale (FN).

To be fair, the Beretta only narrowly beat the popular SIG P226 and was awarded the military contract mainly for reasons of cost. Given the SIG's strong reviews and popularity, I don't think we can reasonably ignore it as a candidate for our apocalypse sidearm. Famous for their top-of-the-line quality and first-rate performance, any of the SIG Sauer pistols should make an excellent sidearm choice for someone arming for the apocalypse.

My first experience with the Beretta pistol was during my last year in the U.S. Army, in 1988, after transferring to an Air Cavalry unit. As a crewmember of a Huey helicopter, I was issued the new pistol and required to qualify with it while wearing the M17 protective mask (gas mask), and I carried this pistol on all our parachute jumps and field training exercises for the remainder of my service. I have to say that I formed a positive opinion of the weapon right away.

As far as how it shoots, the Beretta will hold its own with the majority of comparable automatics on the market. I especially like how easy the Beretta disassembles for cleaning—vastly easier and quicker for me to disassemble than my Springfield Armory 1911. Some shooters will appreciate its ambidextrous safety levers. I remain impressed with the high quality of the product, even as much as I have always liked the 1911A1 pistol that it replaced.

The Beretta M9 is a recoil-operated semiautomatic, double- or single-action pistol of steel construction chambered for 9x19mm Luger/Parabellum. It weighs 33.6 ounces empty and 41 ounces with full magazine (compared with the GLOCK 17 9mm pistol at 22.04 ounces without its magazine and 31.91 ounces with a full magazine) and loads from a 15-round detachable box magazine. This is a large, full-size pistol by modern standards. The Beretta 96 series is the same basic gun chambered for .40 Smith & Wesson.

Taurus International has produced versions of the Beretta pistol, the famous PT 92 and all its variations since Taurus purchased the Beretta factory in Brazil more than 30 years ago. The biggest difference between the guns appears to be that the Taurus pistols have a frame-mounted safety lever, which most shooters seem to prefer, while the Beretta uses a slide-mounted safety lever. Plenty of debate exists over which company makes the better pistol overall, and you will find no shortage of opinions in favor of or against either of them.

Just as with the GLOCK, the Beretta M9 has not been entirely free of controversies and, according to the *Army Times* and other online sources, has been scheduled for official replacement in the near future,

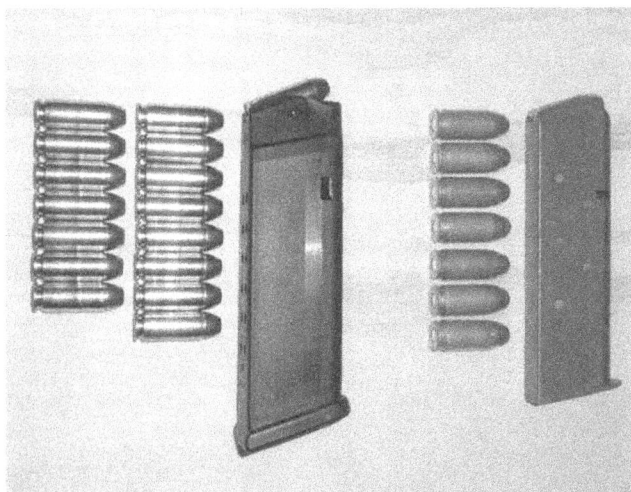

To illustrate the difference in magazine capacity between pistols, the double-column GLOCK 20 magazine on the left holds 15 rounds of 10mm ammunition, while the standard 1911 single-column magazine on the right holds only seven rounds of 45 ACP.

The Springfield XDM 9mm pistol.

depending on the available budget. Nevertheless, the fact that it has remained the primary issue sidearm of the U.S. armed forces for more than 20 years now is something for us to weigh.

A relative newcomer to the auto pistol market to consider are the Springfield XD pistols. These guns were obviously influenced by the GLOCK line, with their polymer frames, but many of the issues we noted with GLOCKs simply do not exist with the XD pistols. XDs have metal guide rods (as opposed to plastic) and fully supported feed ramps, and the XDM (9mm model) comes standard with *19-round* magazines! The GLOCK 17 (full-size 9mm service pistol) has a 17-round magazine. Everything I've ever read about the Springfield XD pistols has been vastly more positive than negative; I cannot even remember one negative point. Give these guns a serious look when considering your apocalypse sidearm.

The XD discussion raises a point we should not ignore when debating the various designs and models of automatics: magazine capacity. The 1911-type pistols, particularly those chambered for .45 ACP (with several exceptions), use single-column magazines that typically hold seven or eight rounds of ammunition. By contrast, the majority of newer generation, full-sized auto pistols feed from staggered magazines holding at least 10 rounds, and many have high-capacity magazines capable of holding as many as 15, 16, or even 17 rounds. This would clearly give any apocalypse survivor a considerable edge in firepower, and

it's something worth thinking about, for those whose local governments haven't yet banned the possession of high-capacity magazines, anyway.

No matter the capacity of your chosen handgun, you would be wise to invest in several extra magazines for it. Magazines are subject to wear and tear—springs wear out, lips get dinged in harsh conditions—and there will be no resupply option after the apocalypse.

I consider question #4 very important: Will our choice of gun and caliber *be sustainable*? In other words, will supplies or sources of spare parts and ammunition be readily available, keeping us operational with it over the long term, both before *and* possibly for years after an apocalypse?

Concerning ammunition choices—and we'll talk quite a bit more about this in chapter 6—it has long been popular among preppers to select personal firearms chambered for common military rounds. The logic follows that surplus military ammunition, especially standard NATO ammunition, could become available to survivors in potentially greater quantity (and therefore possibly cheaper) than commercial/sporting ammunition, and over potentially greater geographical areas and jurisdictions of the world.

Additionally, there is a notion, whether valid or not, that any standard-issue ammunition tested by and considered serviceable for military forces would conform to the highest of standards and should therefore be plenty adequate for most civilian or nonmilitary survival purposes.

Where this line of reasoning might fail us is with the construction of the bullets. Standard military rounds of the world tend to use full-metal-jacketed bullets that don't expand on impact to the same degree as soft-nosed lead and jacketed hollow-pointed bullets. For this reason they may not always be the best for certain hunting or defensive situations, where the mushrooming expansion of a bullet is desirable to create a larger wound channel, as well as to transmit more of the bullet's kinetic energy directly to the target.

The 9mm is the most popular military and police pistol cartridge in the world. Yet the 9mm has always been a controversial pistol round, and not everyone is entirely convinced of its stopping power. I was curious to find out how well a 115-grain +P hollow-point round would punch through 2 x 4-inch lumber, so I conducted two tests.

For the first test, I stacked two short lengths of 2 x 4 framing studs flatways and separated by 1-inch spacers atop a section of treated 4 x 4-inch lumber as a backstop over the concrete floor. I fired the round straight down over the center of the top board using my GLOCK 19 with 4-inch barrel.

The bullet easily passed completely through the first 2 x 4 and came very close to exiting the second but was stopped by a hard knot in that board (see top photo). The bulge at the backside of the second 2 x 4 was nevertheless sufficient to noticeably dent the treated 4 x 4 directly under it, and that second 2 x 4 was also split in half lengthwise in the process, exposing the bullet and what it did to the knot in the board.

For the second test, I duct-taped two pieces of 2 x 4 together, one lying flat against the other, and set this 2 x 4 sandwich over another section of 2 x 4 that was to serve as the backstop. This time I was mindful to avoid shooting into any knots in the wood.

This time the bullet passed completely through both taped 2 x 4 boards and almost exited the back side of the backstop 2 x 4 but was prevented from doing so by the hard concrete floor. The flattened bullet was visible on the bottom side of the third board where it almost exited, but fortunately the floor was not cracked. Had I expected that degree of penetration, I would have used that thicker 4 x 4 piece again as the backstop.

Despite the lack of enthusiasm some pistol shooters have for the 9mm, it takes an impressive amount of force to drive a hollow-point bullet through two or three 2 x 4s. I now believe that the 9mm, especially with a +P load (i.e., loaded to a higher pressure to

Penetration of a 9mm hollow-point bullet in two 2 x 4s. The bullet was lodged in a knot in the second board.

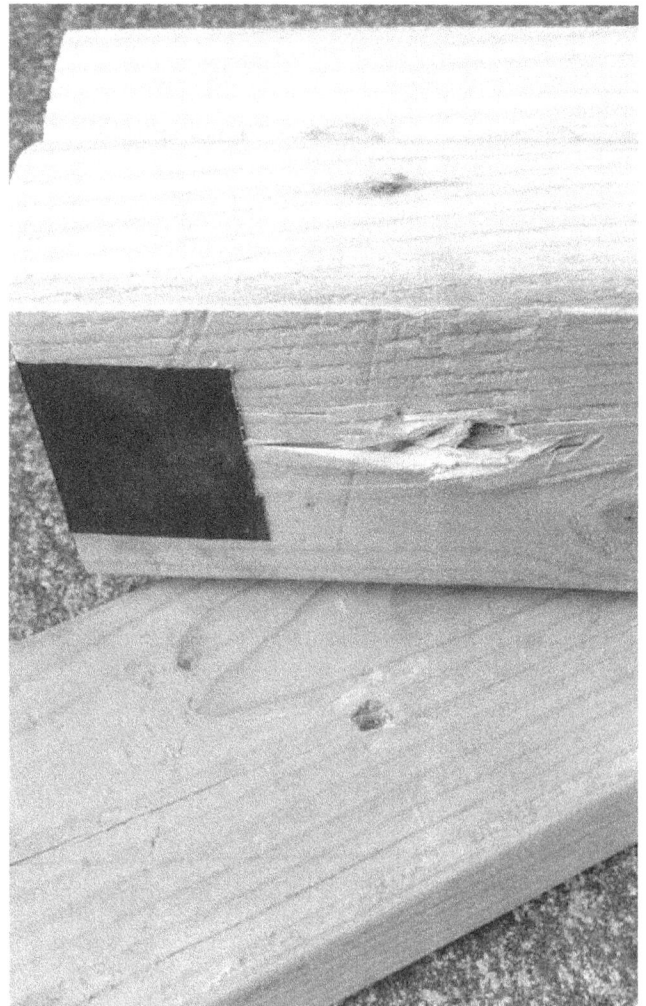

Three 2 x 4 boards penetrated by the 9mm hollow point.

increase muzzle velocity), is a viable handgun cartridge for the coming apocalypse.

The NATO military round for handguns (and numerous submachine guns) at present is the 9x19mm, which is identical (dimensionally anyway) to the 9mm Parabellum/9mm Luger cartridge. Although a distinction is sometimes made between the 9mm NATO and 9mm Parabellum, the only difference is in the load and its chamber pressure standards. The NATO round is

The original 9mm pistol, the P-08 German Luger, is an intriguing classic firearm, but today we have a wide variety of handguns chambered for 9mm with many improved features from which to choose for the apocalypse.

loaded with a 124-grain full-metal-jacketed bullet to a slightly higher pressure in many cases (36,500 psi limit rating in the international CIP standard[1]) than the SAAMI[2] limit rating of 35,001 psi for the 9mm Parabellum. The NATO load is in actuality slightly closer in its chamber pressure standards to many +P loads for the 9mm (38,500 psi SAAMI limit rating for 9mm Parabellum +P). The important thing for us to understand is that modern service pistols chambered for 9mm Parabellum are built to also handle the 9mm NATO round.

A number of 9mm pistol cartridges have been more common in other parts of the world or never became quite as popular as the 9x19mm. The list includes the 9mm Browning Long, 9mm Steyr, 9mm Bayard Long, 9x18mm or 9mm Russian Makarov (recently growing in popularity in the United States), 9mm Action Express, and a few others, but here we will focus our discussion on the more universally popular 9mm Parabellum/Luger.

There is probably no way to verify this, and I admit that it is pure speculation on my part, but if I were to entertain a guess as to which centerfire cartridge (so we can eliminate the .22 Long Rifle rimfire from this pool) has been produced and consumed in greater quantities than all others worldwide, to include all rifle, handgun, and shotgun centerfire cartridges since the metallic cartridge was invented, I would assert that it would have

An Uzi carbine chambered for the 9mm Parabellum.

to be the 9mm Luger. This cartridge has now been in existence and in constant use for over 100 years (since 1908), has served more police and military organizations in Western countries than any other single handgun cartridge, and has been chambered in more models of handguns, submachine guns, and tactical carbines than I could ever hope to list here.

This universal popularity of the 9mm might not be sufficient reason by itself to base our decision for handgun selection, but it is certainly a point worthy of consideration. There is, and likely will be into the foreseeable future, *a lot* of this ammunition floating around.

As far as spare parts availability is concerned, the logical rule of thumb is to go with the most prolific gun models that have been around for a while and have a well-established parts supply infrastructure. Police and military weapons will automatically tend to have this advantage over the majority of sporting firearms. Spare parts will be easier to find for most of the more generic 1911-type pistols, the GLOCK pistols, and the Beretta M9/M92SF than for a lot of other models, given their long run of standardization and widespread use.

And finally let's get to question #5: *how affordable* is our particular choice going to be for us?

Like so many other things, new guns and even many older guns have increased in value with inflation over the years. But the bottom line for most of us is that no matter how suitable or perfect one particular type of weapon is for our purposes, we will ultimately only be able to arm ourselves with whatever we can afford. Survivors will also have to budget for other tools and equipment besides weapons and will be wise to make the most efficient use of personal resources.

For me, this alone rules out most custom-built guns that typically cost in the thousands of dollars (rather than in the hundreds of dollars, as do so many popular factory-made firearms). For example, if I can purchase a new GLOCK or a Beretta 92FS for $550, or even a new 1911 of one brand or another for $600 to $800 (typical prices I see in 2011), then that savings of $1,200 to $1,450 that would have otherwise gone toward a fancy custom $2,000 pistol could instead be used for other goodies like binoculars, backpacks, survival food, hiking boots, knives, tools, radios, gas masks, body armor, flashlights, night-vision optics, reference books, or even that much more ammunition or spare parts for those more affordable firearms.

IF WE SELECT A REVOLVER

The five basic questions we applied to choosing an automatic pistol equally apply to revolvers. We will touch upon all five factors as we weigh the pros and cons of various revolver types and options below.

Revolvers may not seem as alluring as the many "tactical" automatic pistols on the market today, but some people will prefer their feel and handling characteristics over the autos. Shooters who are new to firearms, or wives or older children who will necessarily have greater access to the family guns in a postapocalypse world, may find them more intuitive to load and handle.

As a general rule, auto pistols simply do not lend themselves to hand loading the same way revolvers do. Unlike with revolvers, the operation of an auto pistol typically relies on cartridges operating within a narrower pressure range, and fired empty cases are thrown about when ejected from auto pistols, sometimes making them difficult to gather up for reloading, especially in the brush. Again, at some point in a postapocalypse future, hand loads may be all that's available after cached supplies of conventional ammo run dry.

We can easily see where using an automatic makes sense—when situations are more tactical, or where threats to our safety will most likely come from other armed humans, especially urban street gangs or other postapocalypse hostile groups, rather than from dangerous animals or individual human aggressors. But when the expected self-defense scenario involves a man-killer bear or a bull moose on a rampage in the north woods, you might prefer more muscle with that first shot or two.

This is where a big-bore magnum revolver will have a decided edge over the majority of automatic pistols, simply because it fires a more powerful cartridge. When the threat to our lives will most likely come from some large and dangerous beast, we will want our bullets to impact with the maximum amount of kinetic energy possible.

The .44 Remington Magnum is probably the most popular for this sort of application, but there are others in the same category and even much bigger and more powerful ones nowadays (.500 Smith & Wesson, for one example) that would likely serve just as well or better. For a quick ballpark comparison, the .44 Mag typically provides around 800 ft-lbs. of muzzle energy, while the gargantuan .500 runs in the 1,400 to 2,400+ ft-lbs. range.

One of the main considerations with this big-bore

A hand cannon for the apocalypse—"It could even blow your head clean off!"

The .44 Magnum cartridge on the left, long considered a handgun heavyweight, is utterly dwarfed by the .500 S&W round next to it.

Two revolvers chambered for .44 Magnum. The Taurus Tracker five-shot double-action with 4-inch barrel at bottom left is compact for a big-bore—ideal for backpackers in bear country. The famous Smith & Wesson Model 29 six-shot double-action with 6½-inch barrel at right has accompanied the author on wilderness adventures since the 1980s. It is nickel-plated and wears grips of oak made by the author—a wonderful sidearm, but also cumbersome.

category of revolver is the severity of the recoil. Shooters of small stature and especially inexperienced shooters will find this issue more challenging than others, but even more experienced shooters can develop an undesirable jittery shooting style after touching off enough hand-hammering big-bore rounds.

Another valid concern is the size of the gun, as this affects its portability. The most powerful handguns available—and here we are talking about the truly massive models like the newer X-frame Smith & Wesson revolvers chambered for .460 or .500 S&W, the Ruger Redhawk and Super Blackhawk revolvers, the huge Desert Eagle automatics that fire magnum revolver ammunition, and to some degree even the Smith & Wesson N-frame revolvers like the Model 29—are all comparatively heavy and cumbersome to lug around. The famous Smith & Wesson .44 Magnum from the *Dirty Harry* movies with its 6½-inch barrel weighs 48 ounces—almost equal to two GLOCK 17 pistols without ammo! A backpacker, fisherman, or forager may not feel very comfortable packing so much steel around in a belt holster 24 hours a day, and this can have a bearing on a person's readiness.

Some firearms manufacturers have addressed this issue and developed big-bore handguns that are more compact. Taurus, for example, introduced its Tracker revolvers in .41 and .44 Magnum a few years ago, and these are five-shot double-action revolvers with compact "medium" frames and 4-inch barrels. Smith & Wesson has also stepped up to the plate with their M329 Night Guard, a lightweight and compact six-shot .44 Magnum double-action revolver based on their well-established Model 29, but with a 2½-inch barrel, rounded synthetic grip, and scandium alloy frame to create a handgun that weighs less than 30 ounces.

Moving down the power scale a bit, we find the .357 Magnum, a revolver cartridge that has been without a doubt one of the most popular of the past 70 years.

Any revolver chambered for .357 Magnum has the advantage of being able to also digest all loads of .38 Special ammunition (but the reverse of this is certainly not true), because the magnum case is of the same diameter and only slightly more than a tenth of an inch longer than the Special case. With full-power .357 Magnum loads, the gun will be considerably more powerful than any .38 (the magnum round exiting the revolver's barrel typically in the 1,200 to 1,600 fps range compared with a .38 Special of the same bullet weight at usually something less than 1,000 fps), yet the .357 will still have noticeably less recoil than any magnum with a larger bore, like the .41 or .44 Magnum or heavier rounds.

Even after you decide between single-action and double-action, select the most suitable caliber for your anticipated needs, and choose a particular model or style of revolver, all within a price range that will fit your budget, you will still have the barrel length choice to contend with.

The 2-inch barrel on the typical snub-nosed revolver might be just fine for a house gun or a very close-range defense or backup carry gun, and it will certainly be as concealable as a revolver can be (as we will focus more on in chapter 10), but if the handgun might be called upon to drop a deer at 40 yards in a wilderness survival situation, then a 6-inch or longer barrel would definitely be more practical. A longer barrel has a longer sight radius and thus, as a general rule, it facilitates tighter accuracy than is usually achievable with shorter-barreled guns.

Also, bullet velocity tends to increase with increased barrel length with most ammunition/guns up to a point, sometimes as much as 25 fps or more per inch of revolver barrel length. It depends on a number of

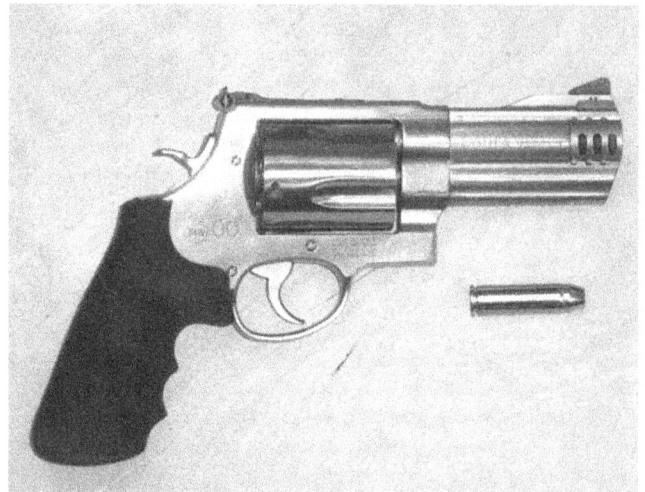

Smith & Wesson's huge X-frame revolver in caliber .500 S&W weighs 56 ounces with 4-inch barrel.

Taking aim with a scoped Ruger Redhawk in .44 Magnum. This is a large revolver that is better suited for hunting than general-purpose carrying.

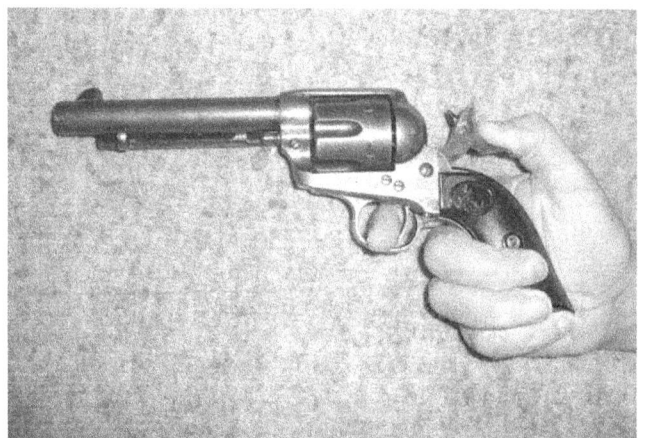

This old Colt Single-Action Army revolver is still tight and very usable, although ammo for this particular .41 caliber is not so common these days.

Loading the swing-out cylinder of a double-action revolver quickly using a speed loader.

variables that can also influence this, from bullet design to the burn rate of the specific powder being used, and even the gun's bore dimensions and tolerances. However, it should also be acknowledged that some hand loaders have reported chronograph results showing higher velocities with *shorter*-barreled revolvers, with all other variables being equal believe it or not, although that is the rare exception to the general rule of increased velocity with increased barrel length.

Since we are discussing revolvers for the apocalypse, we must weigh the single-action designs against the double-actions within this context. For years, I have been a huge fan of the traditional single-action wheel guns that dominated the sidearm realm of the American West, and I do not intend to discourage the reader from selecting any of the great single-action guns—and there are plenty—like the Ruger Blackhawks and Vaqueros and those excellent revolvers

Smith & Wesson's Model 66-3 double-action .357 Magnum 6-shot revolver.

This gun's 6-inch barrel provides longer sight radius that facilitates more refined targeting potential than expected with a shorter barrel.

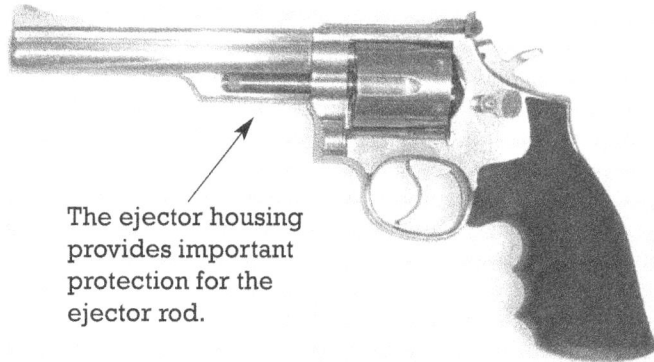

The ejector housing provides important protection for the ejector rod.

The stainless steel resists rust and corrosion better than blued steel.

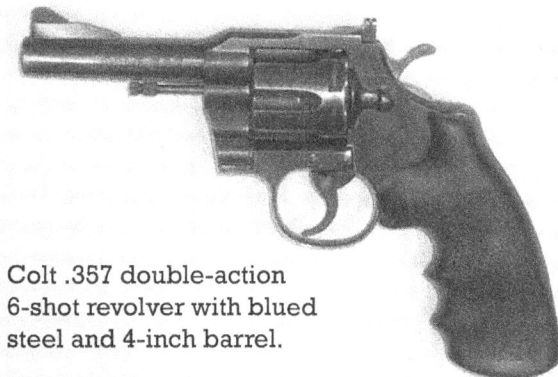

Colt .357 double-action 6-shot revolver with blued steel and 4-inch barrel.

Two excellent .357 Magnum revolvers are shown here to compare features for the apocalypse. The Colt is the author's personal favorite .357 revolver, but the Smith & Wesson gets the vote for the apocalypse scenario.

made by Freedom Arms, to name a few. Any handgun that is powerful enough and accurate enough for the task should be considered a viable contender. Certainly any single-action revolver that is loaded and still shoots could be called upon to effectively save your life in a postapocalypse emergency.

But I think few would dispute the fact that the typical modern double-action revolver—which can normally be fired in either single-action *or* double-action mode—is the more versatile, the more practical in rapid fire, and the faster to reload (due to the double-action's usual swing-out cylinder and its capability of being loaded from speed loaders) of the two basic revolver types.

After weighing all the different features in my own search for the ultimate apocalypse revolver, the closest thing I have found to date is a 6-inch-barreled Smith & Wesson Model 66, which is a double-action, six-shot, stainless-steel revolver chambered for .357 Magnum. Mine wears Hogue soft rubber grips.

There will undoubtedly be readers who will think of other strong candidates for an all-around revolver to meet the requirements discussed in this book (and hopefully so, because it means they're thinking about it). But when we consider the versatility and range of capabilities of the .357 Magnum cartridge (plus all of the .38 Special loads on the market that can also be fired in the gun; Winchester alone lists no fewer than 16 different loads for the .38 Special in its 2012 product guide), the potential speed in which a double-action can be reloaded, the nearly impervious-to-the-elements nature of stainless steel, and the general-purpose practicality of the 6-inch barrel, then this particular gun makes a lot of sense to me.

Again, we face the reality that nothing in life will ever be perfect, and the S&W Model 66 is built on the K frame (the identical twin to the Model 19 S&W, except that the 66 is stainless and the 19 is blued), which is considered by some to be a bit on the light side for the magnum cartridge. This means that a steady diet of the heaviest magnum loads is more apt to loosen the gun's tolerances than would be expected with a frame having more mass, as does Smith & Wesson's N-frame revolvers or any of the beefy Ruger revolver frames for that matter.

On the other hand, a lighter frame does have the advantage of being less burdensome to carry. Toting even one or two fewer ounces in your handgun might free you up to carry several extra rounds of ammunition that could (hypothetically, anyway) save your life at some point in a postapocalypse future.

A WORD ABOUT SIGHTS

When making our handgun selection, the issue of sighting system should be addressed. Options vary from the simplest fixed groove in the frame or V-notch rear with a simple low blade or post front to the most sophisticated telescopic sights specifically made for handguns. Conceivably, your gun could even have no sights at all, as could be the case with an eighteenth century-style horse pistol.

Common today with a lot of medium- and large-sized handguns are adjustable open sights. Sometimes referred to as "target sights" to distinguish them from the simpler fixed, nonadjustable sights, adjustable sights provide a degree of flexibility because they can be adjusted to the specific situation, whether it be shooting at varying distances, using different factory or hand loads with different bullet weights at different velocities that have different points of impact at the target, or simply to facilitate a different sight picture as desired.

Laser sights have also become popular for use on handguns in recent years. With them, the shooter does not align the gun for targeting by viewing the sights on the gun but instead views the projected glowing laser dot image that he positions on the target (assuming the system has been sighted in to match point of bullet impact at the distance intended). This makes target acquisition potentially faster at close range and much easier in dark buildings, and it also allows for accurate shooting from the hip or other positions.

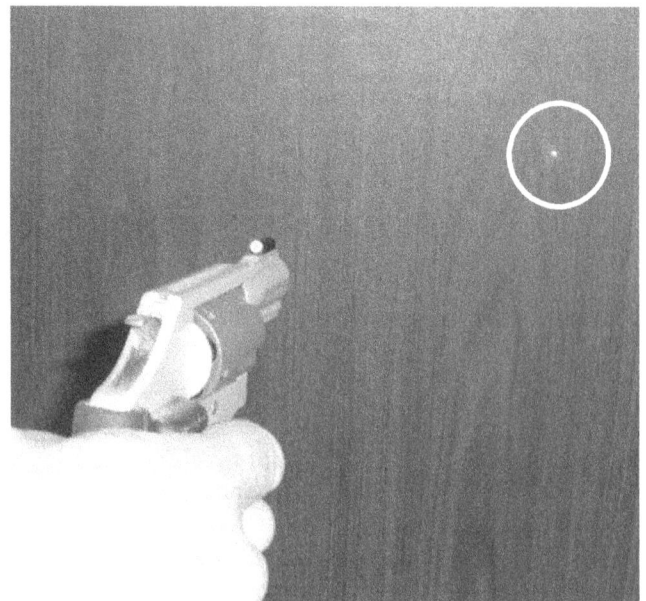

Demonstrating a grip-mounted laser sight on a short-barreled revolver.

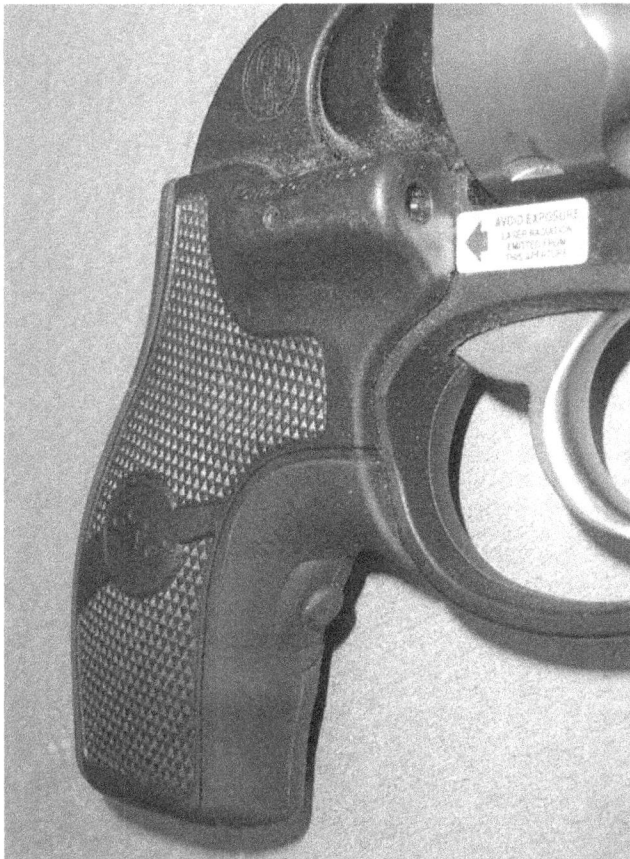

Close-up view of a grip laser sight—Crimson Trace Laser Grips—on a Ruger LCR revolver.

Laser sighting systems have become available in a variety of configurations as the technology has advanced. They can be mounted on the gun's frame, on an accessory rail, under the barrel, or less obtrusively on the grip. There are even inside-the-gun laser units that replace an auto pistol's guide rod.

If there is one drawback to laser sights for use after the apocalypse, it is the battery dependency issue. Even though most of the batteries used for this purpose reportedly have a remarkably long life, eventually batteries go dead. Resupply of fresh batteries likely won't be an option following a breakdown of industrial civilization. If you have a laser sighting system, you need to take this into account as you stock up on supplies.

NOTES

1. CIP, or Commission Internationale Permanente pour l'Epreuve des Armes a Feu Portatives, is an international commission for firearms testing, a proofing organization that sets safety standards for firearms and ammunition for its 14-member countries, mostly in Europe.

2. SAAMI, or Sporting Arms and Ammunition Manufacturers' Institute, is an association of American firearms and ammunition manufacturers that publishes industry standards for their products.

CHAPTER 3

Choosing the Apocalypse Rifle

The two main reasons for using a rifle instead of a handgun are more *accuracy*, especially at extended ranges, and more *power*. When these requirements outweigh the convenience of compactness and portability, then a good rifle is the logical choice. The only question for us, then, is *which* rifle or type of rifle should we select?

Rifles, just like other kinds of firearms, exist in a great number of configurations and action types, calibers, sighting systems, and barrel lengths. While we wade through all these options, we must keep in mind the same five questions we applied to our apocalypse handgun selection:

1. Is the rifle easy to operate?
2. Is it powerful enough for its intended roles?
3. Will it be reliable under adverse conditions?
4. Will it be sustainable in terms of access to spare parts and ammo?
5. Is it affordable to acquire?

As appealing as a single-shot rifle might be in a sporting application (one of my own favorite rifles from my collection is a Browning falling-block single-shot in 7mm Remington Magnum that wears a Leupold variable scope), the survivor forced to fend for himself in a harsh and dangerous world might be better served by one of the repeating rifles available in our time. The ability to deliver rapid follow-up shots could be essential for putting meat on the table or even life saving under the hairiest of desperate scenarios.

Perhaps one of the first questions to be answered in this initial stage of our narrowing-down process is whether you should choose a lever-action, pump-action, bolt-action, or any of the popular semiautos available.

A notion held by a surprising number of people I've talked with is that an automatic encourages expending more ammunition than necessary due to the weapon's rapid-fire capability and the convenience of firing a series of shots.

I have never completely followed this line of thinking, because in my view, a shooter should learn and practice good shooting habits and discipline regardless of what type of weapon he or she happens to be using. Having a high-capacity magazine full of live ammunition at the ready for whatever situation I may encounter does not mean that I have to waste bullets. What it means to me instead is that I simply have more firepower *available* to drop that deer or save my hide in a firefight.

Think about this for a moment—just because our cars are capable of going faster than 100 mph doesn't mean we automatically feel the temptation to drive that fast. It brings to mind what my mom used to say whenever I attempted to leave the house in winter without a coat or sweater: "It is better to have it and not need it, than to need it and not have it."

THE GARAND RIFLES

Given the utility of being able to deliver multiple rounds in quick succession, a semiautomatic rifle seems quite fitting as an end-of-the-world firearm. And if we approach our discussion of semiautos

The M1 Garand service rifle with a bandoleer full of ammunition.

chronologically, our attention is drawn to the classic M1 service rifle of World War II fame, designed by and perfected in the 1930s by John C. Garand, commonly called the M1 Garand.

The M1 rifle holds the distinction of being the first semiautomatic rifle adopted by any major world power as its general-issue infantry rifle, when it was adopted by the United States back in 1936. When the world went to war, the Germans took their Mausers, the Brits their Lee-Enfields, the Italians their Carcanos, the French their Lebels, and the Japanese their Arisakas, and all of those weapons were of the slower bolt-action type. The M1 is still widely regarded as a rifle that was ahead of its time.

Because it has been such a popular weapon for such a long time, much has been written over the years about the M1, but here we will consider some of its main features pertinent to our discussion.

The M1 is a gas-operated semiautomatic battle rifle chambered for (with few exceptions) .30-06 Springfield. It loads from the top of the receiver and uses eight-round clips, which are automatically ejected from the weapon with the last round fired. The weight of the rifle is approximately 10 pounds empty. It features quick-adjustable peep rear sights and protected front post sights.

Some of the weapon's strengths include its robust design and overall ruggedness, functional reliability, famous potential for accuracy (it has been a popular rifle with match shooters for a long time), convenience of operation, and the well-established effectiveness, versatility, and popularity of its cartridge. I personally

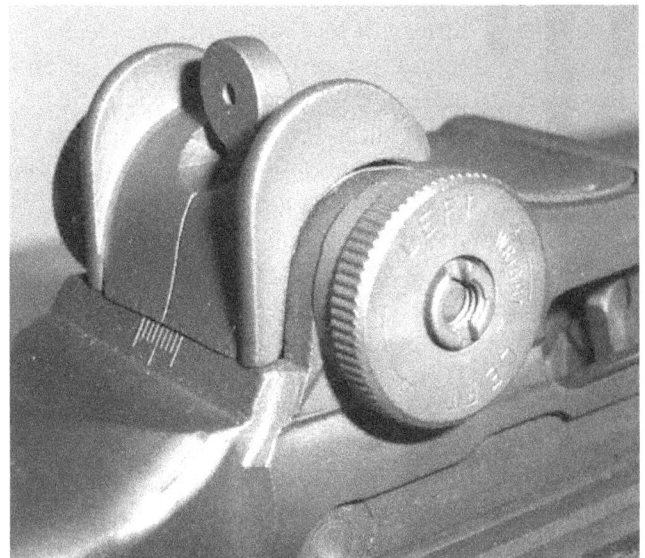

Close-up of the well-protected, quick-adjustable, rear aperture sight on the M1 Garand.

really like the position of the M1-type safety, which is mounted in the trigger guard and conveniently (and quickly) activated by the trigger finger.

The M1's most commonly described disadvantages seem to include 1) its burdensome weight and bulk, 2) its limited (by today's battle rifle standards) eight-round cartridge capacity, and 3) the characteristics of its unique loading/charging system.

Indeed, the way the M1 loads—with those eight-round en-bloc clips into the top of the receiver as opposed to a detachable box magazine underneath, as is more common with contemporary battle rifles—is

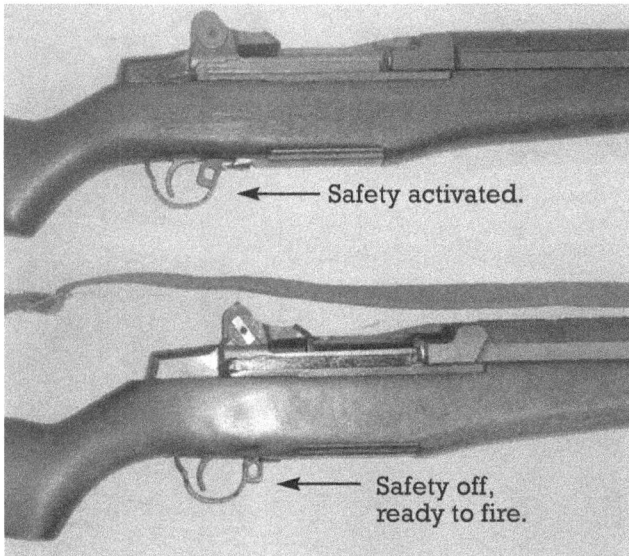

The M1's safety shown here switched on and off.

Safety activated.

Safety off, ready to fire.

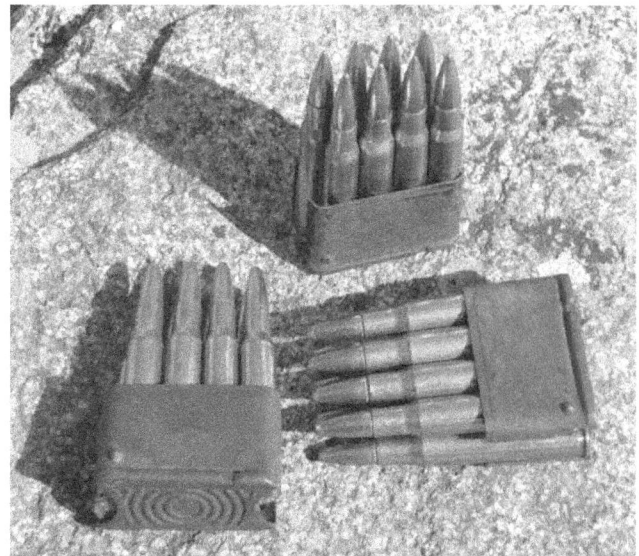

M1 clips hold eight rounds of .30-06.

The M1 Garand loads from the top via eight-round clips.

the particular focus of criticism by some shooters nowadays. The primary gripes are that the weapon won't conveniently accept a partial clip load to periodically "top it off" and keep it always fully loaded, and that characteristic "ping" noise when the empty clip is ejected together with that last empty cartridge casing.

Where others see disadvantages to the system, I see certain advantages. First, it avoids having any bulky metal boxes protruding from the belly of the receiver that would make the weapon heavier and in some respects more awkward, as well as less suitable for shooting from a prone position. Another advantage

to the Garand's clip system is the simplicity of the top-reload process, since there is no magazine that must first be manually removed before another loaded one is inserted. Instead, the freshly loaded eight-round clip is simply shoved down into the open receiver and the weapon is ready to shoot, a bit like using a stripper clip but even slightly quicker because the whole loaded M1 clip is inserted without having to manually discard a stripper clip once the weapon is charged. It is true that the thumb used to push the loaded clip into the receiver must be lifted quickly out of the way of the bolt that slams closed under powerful spring thrust to avoid what is known as "M1 thumb," but this takes only a small amount of practice.

Concerning that popular complaint about the characteristic "ping" sound of the ejecting M1 clips that supposedly announce to the enemy during a firefight that your weapon is empty, I wonder just how significant an issue that is in reality.

It is true that those empty clips do have a signature ring as they're ejected from the weapon together with that last empty cartridge casing, but how easy would it be for anyone to actually hear any ping during the chaos and noise of battle, while other weapons are firing and maybe bombs are exploding nearby? Even if the enemy *could* determine with some degree of certainty that one rifle had suddenly expended its eighth round, what about that guy's buddies' weapon? I would expect that some in a squad or platoon (or in our case, a survival group) would always be reloading their rifles while others were still shooting.

Adjustable
peep sight

Stripper clip
guide

Charging
handle

Trigger and
trigger guard

Safety
lever

Magazine
release

20-round box magazine

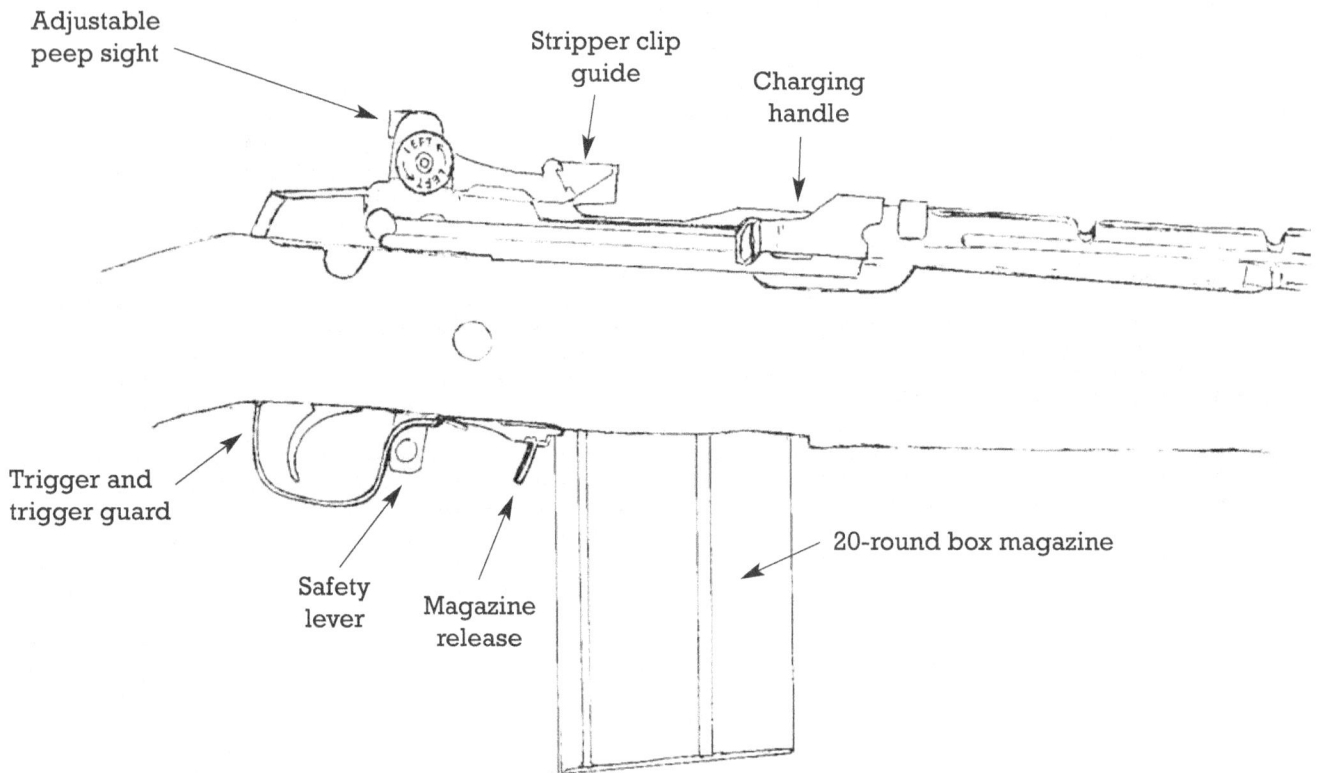

M14 rifle receiver with parts labeled.

But then even if the fight were merely between two individual opponents and one determined that the other had discharged his eighth round from an M1 rifle, the M1 rifleman's reload time with another loaded clip—and we have to assume that he would have more loaded clips in his bandoleer ready to drop in—is mere seconds. You'd also have to assume that the bad guy would be knowledgeable enough about World War II-era small arms to recognize what the ping meant. And he'd have to have the nerve to take immediate advantage of it after having just been on the receiving end of hostile .30-06 rifle fire. Do you see just how insignificant this ping issue is when you look at it rationally?

I find the M1 rifle amazingly easy to shoot accurately with its user-friendly, quick-adjustable military peep sights, and the recoil it produces seems (to me, anyway) to be noticeably less than that of bolt-action sporting rifles chambered for the same .30-06 cartridge or even than the 1903 Springfield bolt-action infantry rifle that it replaced (also chambered for the same cartridge). This is due in large part to its gas-operated

self-loading system that absorbs a certain amount of the weapon's recoil, and also to some degree to its 10+-pound weight (as compared to the typical 8-pound or lighter bolt-action sporting rifle).

Prices for Garands have gone up and down quite a bit in recent decades, and several manufacturers have made M1 rifles, but an average rifle in good condition in 2011 will typically be offered from about $800 to $1,200 in my neck of the woods.

The rifle that evolved from the M1 (and eventually replaced it for service for a brief period at the end of the 1950s and first half of the 1960s) was the M14. The M14 was simply a modernization of a proven weapon, as many of the fundamentals of the M1 esign remained intact.

In redesigning the old workhorse M1 rifle into the eventual M14, the engineers (including Mr. John Garand himself) sought a slightly lighter overall design. The chambering was changed from the .30-06 to the newer (and shorter) 7.62x51mm (.308 Winchester) cartridge, and consequently a shorter and slightly lighter action was thusly made practical.

The M14 can be loaded from the bottom using 20-round detachable box magazines as shown here, or alternatively from the top using 10-round stripper clips (not shown).

The front sight and flash hider of the M14 rifle (bottom) next to the M1 front sight for comparison.

Additionally, engineers added a flash hider to the muzzle end of the barrel, redesigned the gas tube and operating rod configuration, inserted a detachable 20-round box magazine into the belly of the receiver to replace the eight-round top-load clip system, and added a stripper clip guide to the top of the receiver to speed up reloading. The new M14 was also given a select-fire capability so it could be fired as a semiauto weapon like the M1 or switched to full automatic and fired like a machine gun, to serve basically the same role as that of the World War II-era BAR (Browning Automatic Rifle). It has long been a subject of controversy as to how well that full-auto mode actually performed with the .30-caliber infantry rifle, but it's a moot point for our discussion—all civilian versions of the M14, to include the popular Springfield Armory M1A and its variants, are semiautomatic only.

For those wanting to add an M14 rifle to their apocalypse arsenal, there are several possibilities to consider. There have been a number of manufacturers of M14s over the years, and their rifles vary quite a bit in quality. Many of the Chinese-made rifles, for example, have been described as having receivers and bolts of unknown steel alloy and heat treatments with comparatively low Rockwell-C hardness. Additionally, previously demilled receivers have been rewelded and commercially sold as shooters.

I discovered that there is a big difference in quality and reliability between magazines from different makers, and the magazine is a very important component of any autoloading firearm. Before I knew any better, I fancied the idea of those polymer plastic magazines made for M14s, thinking about their desirable lightweight characteristic and resilience to things like shock or corrosion. But after giving several of them a fair test with two different rifles at the range, it became apparent that their plastic followers do not always slide up and down smoothly inside the plastic magazine housing as required but instead tend to get hung up somewhere inside the magazine, repeatedly failing to push the last few rounds up for feeding. I have never experienced anything like this with the GLOCK pistol magazines that are also made of polymer, but I am now convinced that steel magazines with steel followers are the only way to go with an M14 rifle.

I also learned that there is a huge difference in function between the genuine U.S. government-issue M14 magazines and the much cheaper imported steel magazines. At first glance, the imported magazines appear almost the same, but under closer inspection the differences become visible. For one thing, the imported magazines I have seen relied upon a protruded sheet-metal locking lug, while the good mags have a little block of steel affixed to the magazine body to form the lug (see illustration, next page). In my experience, the cheaper sheet-metal version fails to provide firm and reliable locking of the magazine into the magazine well, and the weapon simply refuses to function consistently as a result. Those GI magazines are truly worth the extra money.

M14 magazines: 20-round
The cheap product (left) is unreliable.

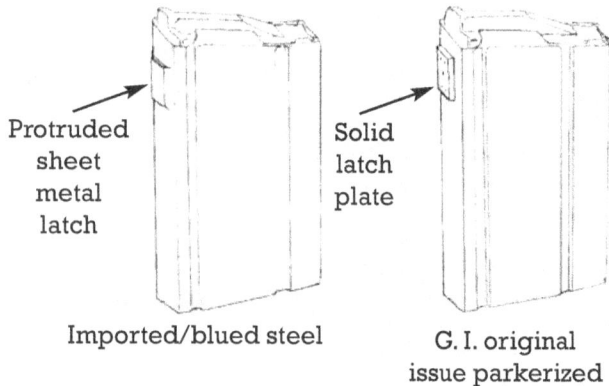

Protruded sheet metal latch

Solid latch plate

Imported/blued steel

G. I. original issue parkerized

▷ M14 magazines.

Springfield Armory has created several variations of its popular M1A rifle in recent years, designating this new generation of M14s the SOCOM series and the Scout Squad rifles. The rifles are offered with several barrel lengths (the Scout Squad has a convenient 18-inch barrel) and various other interesting features like mounting rails for lasers, scopes, or flashlights, as well as muzzle brakes and synthetic stocks. If you are considering an M14 as your doomsday rifle, check out the Springfield website (www.springfield-armory.com). I also recommend three books for more detailed information on M14s: *The M14-Type Rifle: A Shooter's and Collector's Guide* by Joe Poyer, *M14 Rifle History and Development* by Lee Emerson, and *The M14 Owner's Guide and Match Conditioning Instructions* by Scott A. Duff and John M. Miller, with contributing editor David C. Clark.

THE KALASHNIKOV FAMILY OF WEAPONS

The various battle rifles that made their debut across Eastern and Western Europe in the decade following World War II, to include the FN-FAL (and the "inch pattern" L1A1 version), the Mauser-designed H&K G3, and the Kalashnikov AK-47 and its variants, certainly revolutionized the concept of the modern combat rifle. Some of the main features that became standard with the new generation of military rifles and carbines included semiautomatic operation with, in many cases, an optional full-automatic capability

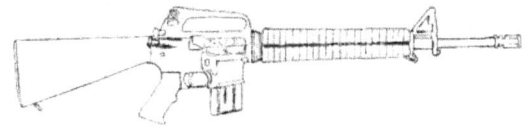

M16A2 Rifle

FN-FAL

H&K Model 91 (G3)

▷ Battle rifles.

(selective fire, like the military M14); box magazines having cartridge capacities of usually 10, 20, or 30 rounds; receiver frames comprised largely of sheet metal; and a more ergonomically configured pistol grip.

The Soviet AK-47 and AK-74; Finnish Valmet M76 and Sako M90; Yugoslav Zastava M70, M76, and M80; Chinese Type 56; and Israeli Galil battle rifles all share the same basic operating system invented at the end of World War II by the Russian small arms designer Mikhail Kalashnikov. In terms of a single modern battle rifle design with the most widespread distribution around the globe over the last 60 or so years, it might be difficult to find anything that comes even close to the Kalashnikov.

My own experience with the AK-47 is limited to one occasion during an Army field exercise at Fort Bragg in 1987, when everyone in our platoon was afforded the opportunity to fire several Russian AK-47s in full-auto mode.

My impression was that the weapon seemed almost crude and awkward in its handling characteristics compared with the M16A1 rifles we were used to at the time. The sight radius and stock of the AK-47 are shorter, making it harder to shoot accurately. Also, the fact that its bolt does not stay open after the last round has been fired the way the bolts in most auto-loading

Romanian semiauto-only AK-47 with aftermarket shoulder stock and rear sight system, caliber 7.62x39 Russian.

A typical AK-47 carbine.

firearms of Western countries do is often viewed as an undesirable design feature. It can slow down the time it takes to switch magazines, as a man in combat would have to count his rounds or hear the click of the firing pin on an empty chamber to know exactly when it was time to reload his weapon.

The AK weapons are definitely not known for their graceful aesthetics or precision quality, but they did establish a reputation early on for being amazingly reliable under all kinds of adverse conditions, and that is a hugely valuable attribute for our purpose of arming for the apocalypse.

THE SKS CARBINE

A similar weapon in some ways is the famous SKS carbine designed by Sergei Gavrilovich Simonov in the early 1940s. I have always thought of the SKS as kind of an economy version of the AK-47, although it is a distinctly different design that lacks the selective-fire capability of the military AK-47s or the convenient quick-change detachable box magazines considered almost standard nowadays. It does use the same cartridge (7.62x39), however, and many of the long list of countries that used the AK-47 in their military forces in the 1950s and 1960s also used the SKS, in some instances as a secondary weapon to the AK.

Prices for SKS carbines have increased significantly in the United States since the early 1990s, when they were normally under a hundred dollars apiece before the Clinton-era restrictions on their importation.

These two Chinese SKS carbines lack the standard-issue swing-out bayonet, but they do sport aftermarket sights and an extended magazine on the bottom gun.

But they still usually run only about half the price of any AK-type weapons that I have seen for sale around where I live. I see a lot of SKS rifles for sale in the $250 to $350 range, and in my opinion they are a reasonable value at that.

AR RIFLES AND CARBINES

Since military rifles historically tend to get smaller and more efficient as new technologies facilitate, it was really only a matter of time until the .30-caliber infantry rifle was made all but obsolete by the space age "black rifle" AR-15/M16 family of weapons and its smaller, lighter, higher velocity (and also controversial) 5.56mm NATO/.223 Remington cartridge. The rifle configuration that emerged was radically different from anything that ever came before it and, as it turned out, formed one of the most ergonomic shoulder weapons ever designed, in my view.

The AR rifles have a straight-line configuration that follows the center of the bore through the whole gun to the butt of the stock. This minimizes barrel lift during rapid fire, reduces felt recoil (not that recoil would be an issue with a .223 rifle), and enhances the aiming properties of the platform. Additionally, the perating controls (magazine-release button, bolt catch, and selector safety/switch) are all within easy reach of either the thumb or index finger of the trigger hand. Finally, the pistol-grip handle is positioned for the most natural hold.

A typical new generation AR-15 carbine in caliber .223 Remington/5.56mm NATO.

Tactical weapons enthusiast Matt Kelso demonstrating the ergonomics of the AR.

These features, combined with an adjustable aperture rear sight (standard, unless set up for a scope or other sighting system), well-protected front post, and the weapon's fast and light low-recoiling cartridge all make for a rifle that is as pleasant to shoot as any rifle ever could be.

I have always considered the AR-15/M16 a high-maintenance rifle, though, because in my experience if the weapon isn't kept exceptionally clean and free of gritty sand, mud, and dirt, malfunctions are likely to occur, probably more likely than with the average automatic weapon. For me, this is a major issue when I am considering a rifle that my very life might depend on.

BOLT-ACTION REPEATERS

As we've been considering thus far, semiautomatic rifles have their distinct advantages for apocalypse survivors forced to defend themselves. However, let's not rule out slower manual-action types.

Bolt-action rifles have the reputation for being inherently rugged, reliable, and low maintenance, but they are also slow. Of the different repeating rifle types, the bolt-action design is perhaps the slowest to operate. The conventional bolt gun requires four motions to cycle its action, while the pump-actions and most of the lever-actions require only two. (The exceptions to this are the various straight-pull bolt-action designs, such as the Lee Straight-Pull or the Swiss Schmidt-Rubin rifles.)

There is a popular perception that the bolt-action is more suitable for precision shooting applications required of snipers, long-range hunters, and bench-rest target shooters. This is mainly owing to the typically tight tolerances in its closed bolt.

The modern bolt-action in at least most of its forms *is* inherently stronger than other repeating rifle action types because of the way the bolt locks up the breech. The strength of the bolt-action is better able to handle cartridges with the highest energy. Because of this, bolt-actions have become dominant with big-game hunters, hand loaders, and serious target shooters, whose passions include envelope-pushing velocities and record-setting ranges.

Even with sturdy bolt rifles, however, hand loaders should use common sense and understand the risks whenever experimenting with the hottest hand loads. Even a super strong bolt-action rifle could blow up under extreme circumstances.

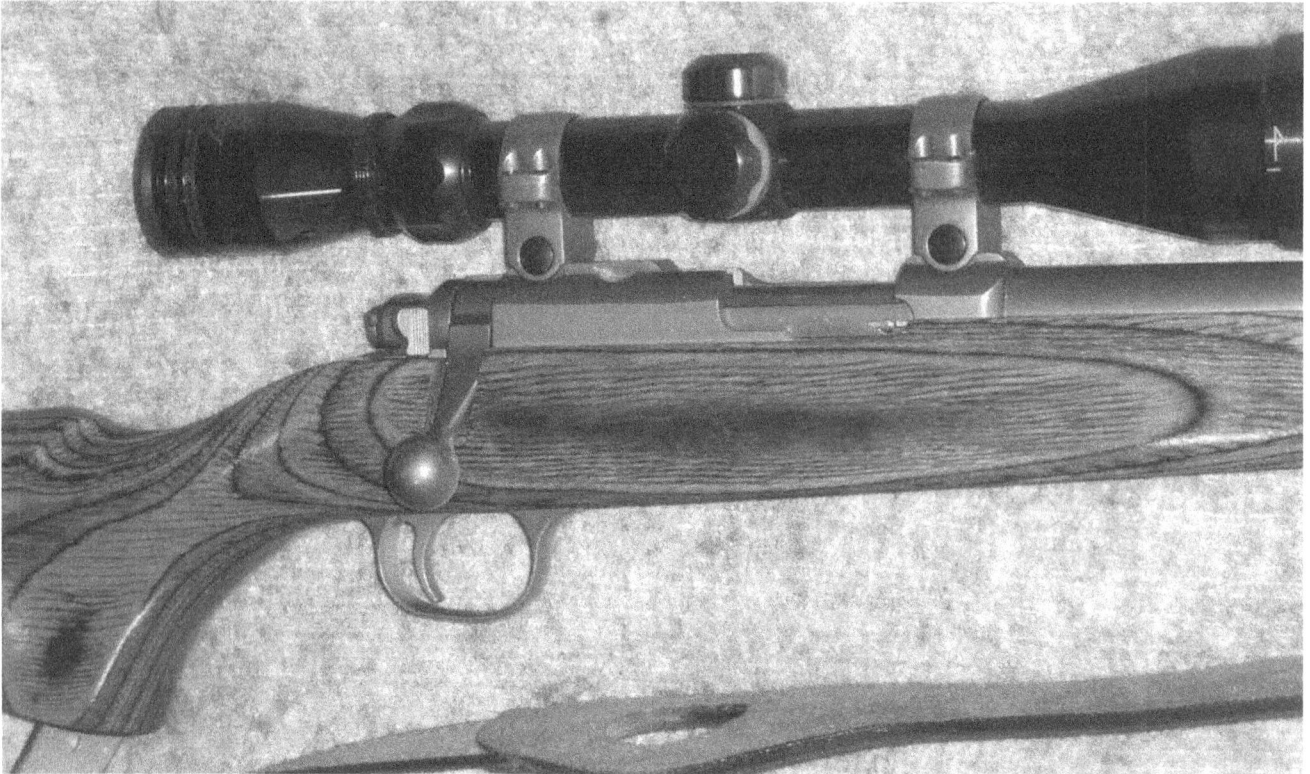

The Ruger Model 77 is a typical modern bolt-action hunting rifle.

A plethora of mostly obsolete bolt-action rifles used during both world wars by almost every major country can still be found at very attractive prices in most parts of the United States. Other than a few of the more collectible models that have increased in value dramatically in past decades (like original unchopped '03 Springfields and the Krag rifles and carbines), a lot of these old service rifles, including many of the Mausers, Lee Enfields, and Mosin Nagants, can still be purchased for less than half the price of the average new sporting rifle.

Military surplus bolt-action rifles, although comparatively heavy and cumbersome, are quite durable as a general rule. Most have a reputation as being rugged and reliable, are potentially very accurate (depending largely upon the condition of their bores and on their sighting configuration), and have plenty of power for 99 percent of what most of us will ever want them for. They were designed and built with the harsh conditions of war in mind, and even if a bit ugly when compared to the finest sporting rifles, they can usually dish it out and take it.

Rugged old military bolt-action rifles from left to right: Pattern 14 Enfield in .303 British, Model 1903 Springfield in .30-06, Krag model 1898 in .30-40, Mosin Nagant 7.62x54R, and Lee Enfield SMLE in .303 British.

THE LEGENDARY LEVER ACTION

Both pump-action and lever-action rifle designs have been around since before automobiles, and both are well-established, proven systems. They use just two movements to cycle their actions, including extracting and ejecting the fired empty cases, feeding the next live round from the magazine into the chamber, and cocking the weapon to make it ready for each shot. The lever-action is popularly associated with the American West during the decades following the Civil War, and I cannot think of any other kind of firearm that seems more fundamentally American than the stereotypical lever-action rifle.

The relatively compact, lightweight, and fast-handling .30-30 lever-action rifle/carbine has been around for more than a century and has certainly taken its share of deer and other game. If this style of firearm is to be considered as the apocalypse rifle, several options must be weighed. The first issue is which brand to acquire, because the two major candidates, Winchester and Marlin, have each sold millions of units and have offered guns with distinctly different design features.

In terms of their general capabilities, there may not be a huge difference between the popular Winchester Model 94 lever-action rifle/carbine and the similar Marlin 336. Both are popular in .30-30 caliber (and have been available in other calibers), have standard open rear sights, weigh almost the same (6½ pounds for the Winchester, 7 pounds for the Marlin), have 20-inch barrels, feature six-round full-length tubular magazines, and manually load and cycle in exactly the same way. The Winchester measures 37¾ inches overall, while the Marlin has an overall length of 38½ inches.

One of the main differences between the Marlin and Winchester is the way they eject empty cases. Winchesters eject the spent cases from the top of the receiver (later Model 94s feature an angle eject), while the Marlin ejects out the right side. This feature makes the Marlin more suitable for mounting a telescopic sight on the receiver (for anyone who actually considers a .30-30 lever-action suitable for scope mounting), although Winchester's newer angle eject was designed to make a telescopic sight more feasible.

The Marlin 336 features a curved lever and pistol grip, while the Winchester Model 94 is standard with a straight wrist and lever. I cannot tell any major difference in handling between these two configurations.

Although lever-action designs that feed from box

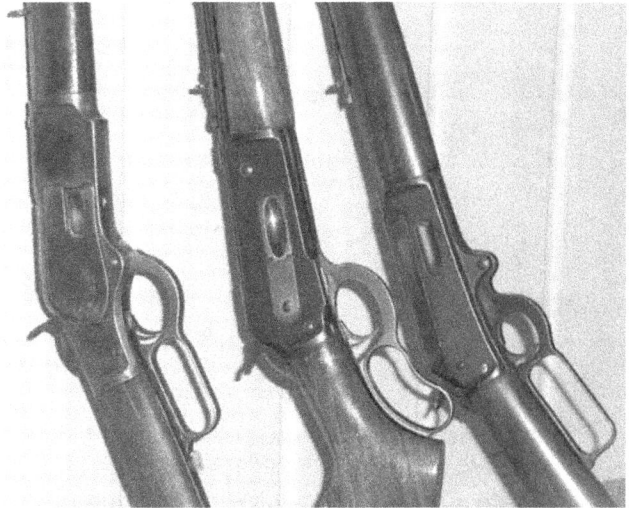

Three classic lever-action rifles, from left: 1873 Winchester, Model 71 Winchester, and Model 1893 Marlin.

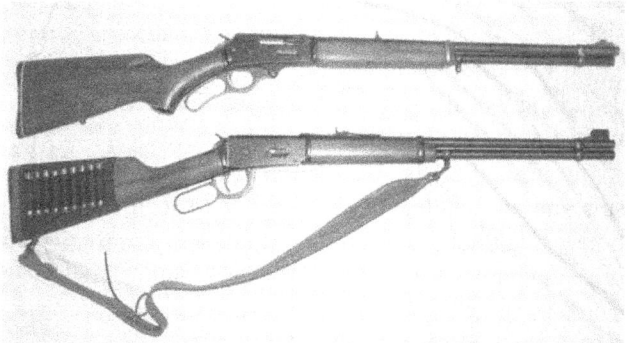

The Marlin 336 at top and the Winchester Model 94 at bottom. Both of these guns are chambered for .30-30.

magazines are not all that uncommon (examples include the Winchester Model 1895, the classic Savage 99, and the currently manufactured Browning BLR rifles), no doubt the majority of available lever-action rifles feed from tubular magazines positioned under the gun's barrel. Advantages to this tube system include greater magazine capacity in many (but not all) instances, especially with full-length magazine tubes on any of the lever guns chambered for such shorter revolver ammunition as .357 Mag, .44 Mag, .45 Long Colt, .38-40, or .44-40. Another advantage is the ease with which the shooter can top off the magazine, because fresh live rounds can be inserted into the side loading port periodically to keep the magazine full without having to open the action or remove a magazine box.

Disadvantages to the tubular magazine system include the vulnerability of the exposed thin steel tube to

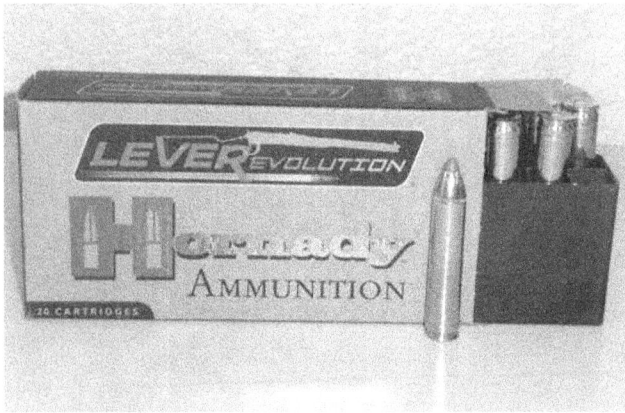

These plastic-tipped, spire-pointed bullets from Hornady are made to safely load and fire in rifles with tubular magazines—in this example caliber .450 Marlin.

Two Winchester Model 94 butt plates. The steel pre-64 plate on the left was made in 1953, the plastic plate on the right in 1981.

Buckhorn rear sights on Model 94 rifles—pre-64 at left, later version at right.

crushing or denting, and the potentially variable compression force—and consequently the occasional inconsistency—of the long, lightweight coil magazine spring.

It was traditionally recommended that ammunition used in any of the lever- or pump-action rifles having magazine tubes should have blunt, soft lead, flat- or round-nosed bullets. This was because copper-jacketed, spire-pointed bullets could be prone to accidental ignition inside the tube, where the tip of one live round presses against the primer of the cartridge in front of it. The concern was that the inertia of recoil could simulate the action of a firing pin striking the primer.

This was always an annoying limitation of the tubular magazine guns for a lot of long-range hunters, because blunt-shaped bullets are simply not as aerodynamic as spire points. Consequently, lever-action rifles with tube magazines have traditionally been thought of as strictly thick timber or brush guns. Hornady's new LEVERevolution bullets effectively remove this limitation; they feature soft plastic tips in the noses of aerodynamic spire-pointed bullets, making them safe for use in tubular magazines.

The Winchester Model 94: Pre-64 vs. Later Variants

For those who prefer the style of the classic Winchester Model 94 to other contemporary lever-action guns, it might be worth knowing—without getting into the many variations of this basic model made before World War II known as "pre-war"—that there are differences in quality between the Winchesters manufactured before 1964 and those made after that pivotal year, when cost-cutting changes were implemented into the production. Many of the steel parts that were

previously machined from forged stock were replaced with cheaper cast and stamped parts, steel butt plates were replaced with plates made of hard plastic, and, at least with the examples I have seen, the traditional buckhorn-style rear sight was made noticeably thinner and simpler.

I have firsthand experience with the inferiority of the cast-steel parts in the late Model 94s. About a year after I purchased my Model 94 Trapper carbine in .44 Magnum (new in 2002), I decided I wanted to sharpen my skills cycling the action with maximum swiftness (Chuck Connors style), and I started practicing my rapid lever-action techniques with the new gun. It wasn't long before the thin rail sections on *both sides* of the link—being cast steel in this gun—broke off from the repeated stresses of this activity, rendering the action completely inoperable.

Fortunately for me, my dad just happened to have several extra links from older Model 1894 rifles in his box of Winchester parts (his passion is tinkering on the old guns), and with some modification we were able to adapt one of those older parts for use in my newer gun. That was almost nine years ago, and this little carbine has functioned flawlessly ever since. I mention this situation because anyone who is shopping for weapons for the apocalypse will want to acquire the most durable products available to last them for the longest amount of time.

So look for Winchester Model 94s having serial numbers below 2,600,000, which means they were manufactured prior to 1964. Not only are they of better quality as just noted, but generally they are worth more money than the later Model 94s.

Takedown Lever-Action Rifles

Lever-action rifles designed such that their barrels can be quickly separated from their receivers have been moderately popular in years past, because in their disassembled state their overall length is much shorter, making them easier to store in confined spaces or transport in a backpack or duffle bag.

As of this writing, Browning offers its popular BLR rifles with the handy takedown feature (www.browning.com). The BLR feeds from a box magazine, so there is no tube to fiddle with during the takedown procedure. What makes these modern Browning lever actions especially appealing to me is that they are available in 15 different rifle calibers, from .223 Remington to .450 Marlin, and even .300 Winchester Magnum!

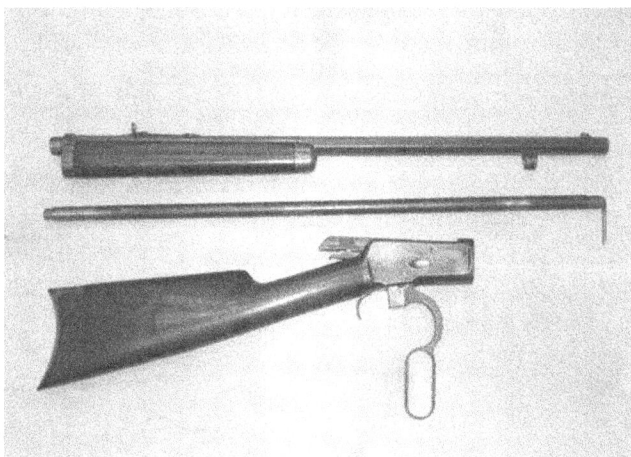

Unscrewing the magazine tube to take down this rifle.

CARBINES AND THE SCOUT RIFLE CONCEPT

If portability or maneuverability take priority over maximum effective range or even precision long-range accuracy, the shorter barrels and the lighter weight of carbines make them attractive. Carbine versions of lever-action rifles, bolt-actions, single-shots, and semi-automatics having barrel lengths from 16 to 20 inches have all been popular.

Shoulder weapons of carbine size are often chambered in handgun cartridges. Their shorter barrels are still long enough to facilitate complete burn of the handgun round's powder charge, which may not always be the case with some of the large, powerful rifle cartridges when fired from barrels shorter than 18 inches. Also, the lighter weight of the carbines can make them punishing on the shoulder with the level of

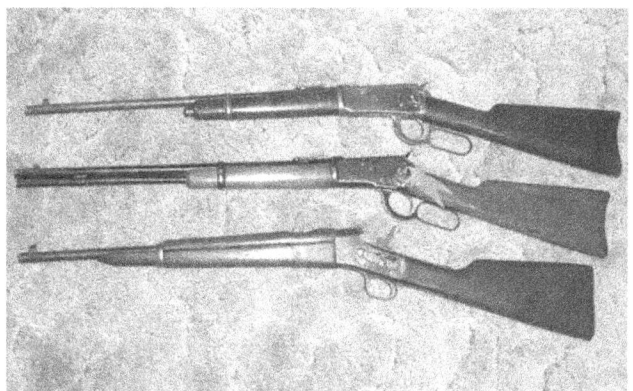

An old .38-40 Winchester Model 1892 rifle with the rare takedown feature, shown disassembled into its three components: action and stock, barrel, and magazine tube.

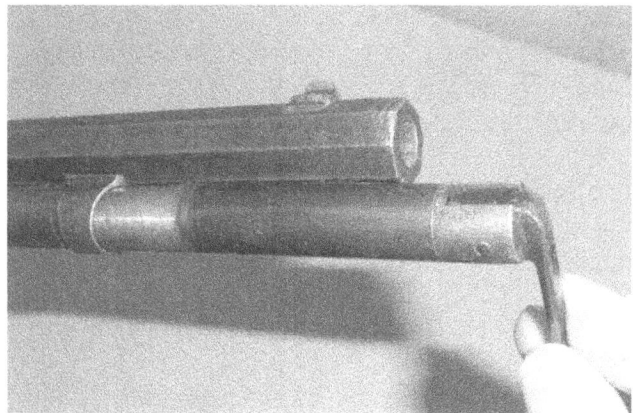

Typical "saddle-ring" carbines from the late 1800s. Being light and short, these were especially convenient for cavalry soldiers, Indians, and cowboys who spent a lot of time on horseback. They might also be practical for future apocalypse survivors.

recoil generated by some of the larger rifle cartridges, and this is usually not the case with the lower-energy handgun rounds.

But even as conveniently utilitarian as the handgun cartridge in the carbine arrangement can be, handgun cartridges cannot generally compete with most of the common rifle cartridges when it comes to long-range bullet performance. Our only consolation in this is that a bullet—in this case a handgun bullet—fired from a carbine will outperform the same bullet fired from a handgun.

A comparatively compact shoulder weapon that fires a powerful rifle cartridge definitely has its merits for those who need rifle-like performance from the smallest practical platform. An example of this is the Tanker M1, which is a shortened version of the M1 Garand. A full-size M1 measures 43½ inches in overall length, while the Tanker is only 37¾ inches long. Both versions are commonly chambered for .30-06 Springfield. Such a weapon would surely be a godsend in just about any kind of survival situation. However, the Tanker model is not as common (or as easy to find) as the standard M1 rifle.

The short version of the M1 Garand rifle, the M1 Tanker.

Citizens arming themselves for the coming apocalypse may wish to consider a kind of hybrid between the contemporary full-sized rifle and the smaller carbine called the Scout Rifle, which was originally conceptualized by the famous firearms instructor and writer Jeff Cooper. Variations of this basic theme exist as either bolt action or semiauto, having barrel lengths from 16½ to 18 inches (16 inches being the minimum civilian-legal barrel length for a rifle). These guns are chambered for a rifle cartridge; the most common is perhaps the .308 Winchester. Scout Rifles sport iron sights and usually a Picatinny rail scope mount in the handguard area that positions the optic well forward of the gun's action. The basic Scout Rifle platform is usable with a special scope or its iron sights.

The Scout Squad from Springfield Armory mentioned earlier is a semiautomatic version of the Scout Rifle concept, and Ruger recently introduced its M77 bolt-action version of this configuration chambered for .308 Winchester. Called the Gunsite Scout Rifle, it features a 10-round detachable box magazine, 16½-inch barrel, Picatinny rail for scope mounting forward of the receiver, aperture ghost-ring rear sight, protected front sight post, flash suppressor on the muzzle, and black laminate stock. The total weight is 7 pounds, and the overall length ranges from 38 to 39½ inches, making this unit slightly shorter and more than a pound lighter than Springfield's Scout Squad.

One of the benefits of the Scout's forward-mounted scope is that the shooter never has to be concerned about getting smacked in the face by the eyepiece during recoil—the famous "scope cut" hazard that seems rather common among shooters using scope-sighted rifles in the heavy belted magnums. The eyepiece of the scope on the Scout Rifle is too far forward to present any chance of this happening.

Of course, only a scope with low magnification and an especially long eye relief (or LER) is properly suited for this application. Leupold has a scope intended especially for Scout Rifles, the FX-II, that has 2.5x28 magnification, which they call IER, for "intermediate eye relief." Burris Optics' Scout Scope with 2.75x is another popular telescopic sight specially designed for the Scout Rifle platform.

HOW ABOUT A RIFLE FOR LONG RANGE?

Your ability to maintain distance between you and your adversaries or to reach out long distances to kill

This Remington Model 700 Sendero is a bull-barreled bolt-action rifle designed for long-range shooting. This one is chambered for .300 Win Mag and wears a Bausch & Lomb Elite 4000 variable 4-16x scope with a shade tube.

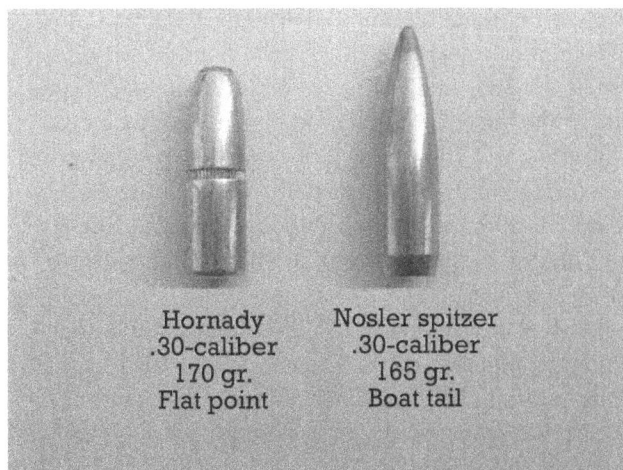

Hornady
.30-caliber
170 gr.
Flat point

Nosler spitzer
.30-caliber
165 gr.
Boat tail

Two different bullet designs for comparison. Assuming the same velocity with both, the tapered boat-tail bullet at right will perform better at long range.

game could potentially be a crucial component of your survival in the apocalypse, so we should at least consider a few possibilities within this realm.

Let's begin this discussion talking about cartridges and projectiles most suitable for the longer shots. (For our purposes, we'll consider any distances greater than about 200 yards, including all the way out to and beyond 1,000 yards, as fitting into the long-range category.) Bullet trajectory will naturally be an important issue to those needing a rifle for taking shots at the longer distances. The objective is to flatten out the arc of the bullet path as much as possible.

The one variable (actually a measure of multiple variables) that will influence this more than anything else is the bullet's ballistic coefficient (BC). The ballistic coefficient is a measure of how well a projectile flies through the air, or the measure of the bullet's ability to overcome air resistance in flight. This has a lot to do with the aerodynamic nature of the projectile, and it factors in such physical characteristics as the projectile's shape and contour, its sectional density, and its velocity.

Rather than bore the readers with a lengthy discussion about ballistic coefficients, I think it is more useful to simply understand that a bullet with a higher BC number will retain more of its velocity and energy, shoot flatter, and consequently perform better down range.

The shape and design of the bullet will significantly affect its BC. The popular boat-tail (BT) feature

of some high-velocity rifle bullet designs has a slightly tapered base that serves to reduce the drag at the back of the bullet and therefore increase its ballistic coefficient. Thirty-caliber rifle bullets in the heavier weight ranges (from around 165 grains on up to around 200 grains) have been popular with military snipers and long-range rifle competitors for decades. Shot through specialized rifles chambered for .30-06, .308 Winchester (7.62x51), and .300 Winchester Magnum, these bullets exhibit comparatively high ballistic coefficients. As was explained by Maj. John L. Plaster, USA (Ret.), in *The Ultimate Sniper* (a very popular Paladin Press book), the .300 Winchester Magnum (talking about the 200-grain boat tail in this case) still has more kinetic energy at 1,000 yards than does a .44 Magnum handgun at point-blank range.

For some comparisons, a .300 Win Mag bullet that weighs 180 grains and has a muzzle velocity of 3,000 fps will normally have a ballistic coefficient higher than .500, while a 170-grain .30-30 bullet—also a 30-caliber, mind you—having a muzzle velocity of 2,200 fps will do well to reach a BC of .245, and even that is considerably higher than the .170 to .180 BC of a typical 110-grain .30 Carbine bullet that exits the muzzle at just under 2,000 fps. So, we can see from all of this that projectiles that are proportionately longer for their diameter, aerodynamically shaped, have high mass, and fly the fastest will have the highest ballistic coefficients.

For an idea about how this affects the trajectory of bullets, we might compare the bullet drop averages of two common deer hunting cartridges, the .30-30 and .30-06. Assuming both rifles are sighted in at 100 yards, the 165- and 168-grain bullets from the ought-six, in the muzzle velocity range of 2,800 fps, or the 180-grain at 2,700 fps, will normally drop somewhere between 13 and 15 inches at 300 yards, while a 170-grain .30-30 bullet that exits the muzzle at 2,200 fps will normally drop between 30 and 32 inches at that same 300 yards.

A few other popular rifle cartridges known for having comparatively flat trajectories up to at least 500 yards are the old .270 Winchester and the newer .270 WSM, the 7mm Remington Magnum and 7mm WSM, and the .300 WSM.

There has been a trend in recent years toward larger and heavier bullets for long-range shooting. Bullets of .338- to .416-inch diameter have become increasingly popular. The .338 Lapua Magnum developed for military long-range sniper rifles, for one example, can be loaded with a 250-grain bullet to 3,000 fps, and the .338-378 Weatherby, for another example (which is actually the .378 Weatherby case necked down to .338 caliber), can be loaded with a 300-grain bullet to a muzzle velocity of over 3,000 fps.

The .300 Win Mag round on the left is dwarfed by that .50 BMG round next to it.

Jerry Diemer hefts his Barrett M82A1 .50-caliber rifle that weighs more than 30 pounds.

For those wanting even more energy down range, the .50-caliber rifles are in a category all their own.

A friend of mine, Jerry Diemer, owns and shoots a Barrett .50-caliber semiautomatic rifle, the famous Model 82A1. This basic rifle and its popular cartridge have served U.S. military snipers in the Middle East for more than two decades now, and this just might be an apocalypse weapon to consider for those who want the absolute maximum amount of power and range capability that a civilian can legally own.

According to Barrett's website, the M82A1 weighs 30.9 lbs and is 57 inches in overall length. (Jerry measured his rifle at 56.25 inches, and it weighs in at 33.2 pounds with its scope and empty magazine.) This is a recoil-operated semiautomatic with a 10-round detachable box magazine. With its .50 BMG (Browning Machine Gun) cartridge, the M82A1 is effective not only against personnel but light vehicles and equipment and has an effective range out to 2,000 yards, or nearly 1 1/6 miles. That's close to twice as far as any .30-caliber rifle can realistically provide useful effectiveness.

Jerry mentioned that a common bullet weight for his gun is 610 grains. In *Cartridges of the World*, 8th edition, by Frank C. Barnes, three different bullet weights are shown for the 50 BMG: 660, 750, and 800 grains. The muzzle velocity shown for the 660-grain bullet—listed as a PMC factory load—is 3,080 fps, with a muzzle energy of 13,910 ft-lbs.! I am not aware of any sporting rifle having that much power.

The perceived recoil of the Model 82A1 is not harsh by most standards, owing to its recoil operation combined with its considerable weight and special muzzle brake. Its straight-line design from the barrel to

Side view of the Barrett's receiver and scope.

Muzzle end of the Barrett rifle.

With its overall weight and attached bipod, the Barrett rifle is ideally suited for shooting from a prone position. Jerry uses machine gun belts as convenient bandoleers for the big .50-caliber rounds, but the Barrett rifle is not a belt-fed weapon.

This heavy custom .30-06 bench-rest rifle wears a fixed-power Leupold target scope having 36x magnification, which is great magnification for long distance but excessive for optical clarity within 100 yards.

the butt is no doubt a contributing factor in reducing felt recoil as well. The recoil is often described as similar to that of a 12-gauge shotgun. I don't remember the Barrett's kick being even that much, although I have to admit it's been about 10 years since I fired Jerry's gun, and I fired only one round.

Due to the rifle's size and substantial weight, together with the weight of its ammunition, the Barrett system is clearly better suited as a long-range rifle fired from a stationary position, as opposed to any of the lighter, more portable rifles used in the sniper role. Also, the Model 82A1 is certainly not a poor man's rifle. With an appropriate scope and supply of ammunition, the cost will be something very close to $10,000.

Modern rifles built for long-range shooting will usually have a full-length (24 to 28 inches) free-floated heavy bull or semi-bull barrel for minimizing vibration and maximizing stability. They will wear a high-end telescopic sight having fairly high magnification or a variable option that allows changes from medium to high power, such as 4x to 16x magnification, and sometimes with built-in range-finding devices. They will often sport an adjustable trigger and a stock custom-fitted to the shooter, and they will be chambered for a cartridge capable of delivering top performance at extended ranges, as might be expected of any of those rounds just mentioned.

Often, a rifle barrel chambered for the highest-energy belted magnums will be ported or feature some type of muzzle brake to help tame recoil. The bolt action is probably the most popular action type, given its characteristic precision. It is not uncommon for such specialized rifles to weigh anywhere from 12 to 16 pounds with their scope, bipod, full magazine, and sling.

I think it is important to remember that these often heavy and awkward long-range rifles are very specialized pieces of equipment that simply cannot be made to fit properly into any kind of general-purpose rifle category. Consider one for your doomsday arsenal only if you have a true, practical need for its unique capabilities.

CHAPTER 4

Choosing the Apocalypse Shotgun

For within its effective range, I cannot think of any civilian-legal weapon that is potentially more devastating than the big-bore shotgun. Multiple pellet wounds from a single shot shell fired at close range will constitute massive tissue damage to either man or beast, putting an end to most kinds of threats in an instant. But this kind of destructive potential is only one of the shotgun's several attributes.

One of the important things to remember about arming for the apocalypse, and I believe this warrants reiterating, is that we will want maximum versatility in our weapons, because survival situations are so varied and unpredictable. And it just so happens that versatility is a major basic trait of the shotgun, because by varying the size of the shot pellets in the loads—or alternatively using individual slugs—the weapon can be made suitable for so many different categories of hunting or for a wide range of specialty, tactical, and defensive purposes.

In the United States today, the 12-gauge is by far the most popular shotgun size, and for this reason more than any other, the "twelve" would be my *first* choice of shotgun for the apocalypse. In fact, if I were pressed to select the single most versatile basic firearm type available today, I would go with the 12-gauge shotgun.

The "scattergun" is an incredibly versatile tool that can shoot individual slugs of various designs, heavy buckshot, and all the different sizes of birdshot. But over the years, the options have increased considerably with so many new developments.

Knut Rogers aims the 12-gauge scattergun at a plywood board 22 yards away.

The saboted slug is just one example of an innovation that has expanded the versatility of the shotgun, allowing a smaller diameter (typically .50-caliber) slug to be launched at higher velocity than is achievable with a plain lead slug of 12-gauge bore diameter. It does this with an inner bullet encased within a plastic sabot that peels away after the projectile exits the muzzle, allowing the inner bullet to fly faster (and with a better ballistic coefficient than that of a regular slug) to the target. Winchester's Dual Bond sabot slug loaded in a 3-inch shell is advertised as having a 375-grain jacketed bullet that will exit the shotgun barrel at 1,850 fps. This configuration is meant to be used in conjunction with a rifled shotgun barrel, however, which is needed to stabilize the bullet. The standard

Showing the back side of the half-inch plywood board ventilated by a 15-count 3-inch Magnum 00 buckshot load.

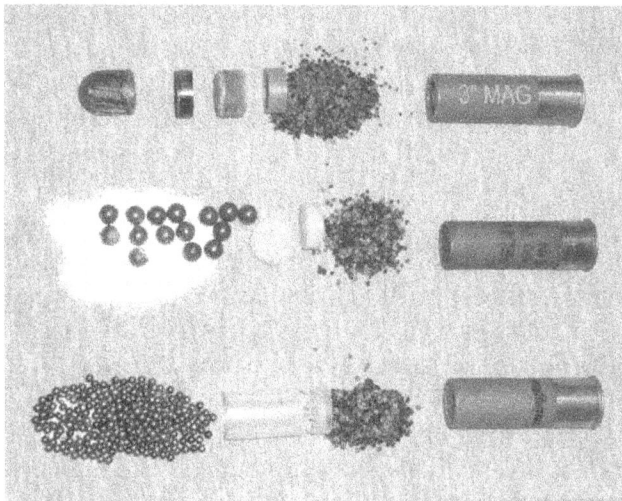

.410 cal.   28 ga.   20 ga.   16 ga.   12 ga.   10 ga.

Shot shells of available gauges, arranged by size.

Close-up of a 12-gauge rifled slug, showing its hollow base.

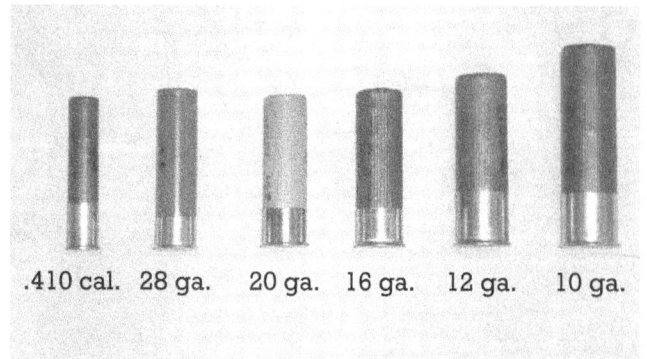

Some 12-gauge shot shells dissected to reveal three very different common shotgun loads: Remington 1-oz. rifled slug at top, Federal Classic 00-Buckshot 3-inch Magnum in the middle that holds 15 pellets, and Remington turkey load at bottom loaded with 2-oz. buffered copper-plated #4 shot.

12-gauge slug commonly used in the United States is the Foster slug, which is a rifled, hollow-base, round-nosed lead projectile intended to be fired in a smooth-bore barrel. Unlike a solid lead projectile such as a round ball, the Foster slug will safely swage down when fired through a choke. Its forward mass configuration helps to stabilize it. The "rifling" fins on the slug impart no spin at all to the projectile but do serve to minimize friction inside the barrel. The usual weight of a 12-gauge Foster slug is 1 ounce (or 437.5 grains)

There is a greater variety of loads on the market for the 12-gauge than for any of the other shotgun gauges simply because it is so much more popular. In reality, any of the shotgun bore sizes have great possibilities for the apocalypse survivor. For a while it almost looked like the old 16-gauge was becoming obsolete, but lately I've seen quite a variety of new 16-gauge loads on the ammunition shelves. The advantage of the 16-gauge for someone arming for

An example of an adjustable choke on the muzzle of a 12-gauge pump shotgun.

Lead Buckshot #	Diameter in inches
4 buck	.24
3 buck	.25
2 buck	.27
1 buck	.30
0 (ought buck)	.32
00 (double-ought buck)	.33
000 (triple-ought buck)	.36

Lead Birdshot #	Diameter in inches
12	.050
9	.080
8½	.085
8	.090
7½	.095
6	.110
5	.120
4	.130
2	.150
BB	.180

Steel Shot #	Diameter in inches
6	.11
5	.12
4	.13
3	.14
2	.15
1	.16
BB	.18
BBB	.19
T	.20
F	.22

Shotgun Bore Diameters (approx. inches)	
10 gauge	.775
12 gauge	.730
16 gauge	.670
20 gauge	.615
28 gauge	.550
410 bore	.410

the apocalypse on a budget is that the guns tend to be, more often than not, lower priced than their equivalents in 12-gauge.

A shooter's aiming requirements with a shotgun are much less critical than with a single-projectile weapon due to the progressively widening pattern of shot that allows a greater margin for aiming error. In this regard, we concern ourselves a lot with these shot patterns when considering shotguns. With any bore configuration, we want the ideal combination of the widest possible spread of pellets to allow for the greatest margin of aiming error, along with a dense enough pattern of pellets to reliably hit with sufficient quantity of shot to kill whatever size game we are hunting, at the most probable hunting or targeting distances.

The barrel's choke—the degree to which the inside bore is constricted toward the muzzle—will influence shot pattern size at a given distance more than any other variable. Constricting the inside of the tube in this way will have the effect of tightening the shot pattern. The major shotgun bore configurations—with several "in-between" variations—are cylinder bore (or no choke at all, i.e., straight tube), improved cylinder (.010 restriction), modified (a moderate choke having .020 restriction), and finally the various full chokes: light full, full, and extra full (.040 restriction), which will produce the tightest shot patterns.

Shot size is also an important factor whenever using shotguns for any purpose, because we want our pellets to be large and heavy enough to reliably kill or in some situations destroy the intended target at the distances we expect target engagement. Here we find birdshot ranging from small (like #9 to #7½) to the larger birdshot up to BB size, plus the various sizes of much heavier buckshot (see accompanying shot size chart).

Shotguns for defensive or tactical purposes will be used with heavy buckshot much more often than with other shot sizes, and probably the most popular is double-ought buckshot (00 buck). A 2¾-inch 12-gauge

The business end of a 12-gauge double-barreled side-by-side.

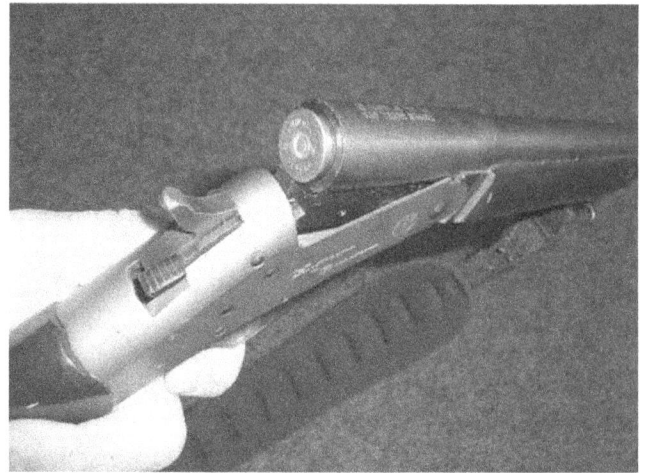
Loading the break-action single-shot gun, as can be seen here, is a very simple process.

shell loaded with 00 buck, including the standard combat loads used by the military, will usually contain nine of the .33-caliber lead round balls, while 3-inch magnums will typically be loaded with 15 of them. A single 00 buckshot weighs very close to 49 grains on my scale, so nine of them will weigh close to 440 grains, and 15 will add up to 735 grains. These heavy buckshot loads typically exit a shotgun barrel from around 1,200 to more than 1,400 fps, so it should be no wonder that they beat up the shoulder. Other buckshot sizes are also popular for home defense and hunting purposes.

The two main *disadvantages* to shotguns, besides the characteristic heavy kick (recoil), for long-term postapocalypse survivors are 1) their limited range, and 2) their comparatively heavy and bulky ammunition. The potential advantages, however, are many, as we've discussed.

A word of warning: do not be tempted to saw a shotgun barrel down to some shorter length to increase its concealability, portability, or shot spread at close range. In order to stay legal in the United States (at least in this pre-apocalypse era), shotgun barrels must be at least 18 inches in length, and the overall shotgun length must be at least 26 inches to conform to the National Firearms Act.

COMMON SHOTGUN ACTION TYPES

Whenever considering a shotgun to serve as our emergency survival tool, we will have to decide between the pump (slide action), the semiauto, the single shot, or the double-barreled (side-by-side or over-under). Other shotgun designs like bolt-actions and

lever-actions have appeared in the past occasionally, but they are less common today.

A double-barreled shotgun has the advantage of having two different chokes with the same gun. It is not uncommon with bird guns to have one of the two barrels choked with either improved cylinder or modified and the other barrel with a full choke. My dad's LC Smith 12-gauge side-by-side is arranged this way—one barrel is modified and the other has a full choke. Thus, a hunter can fire the first barrel with the wider spread at the closest bird and the tighter choked barrel at the farthest bird, and his shot patterns at the different distances would be well suited for the task.

Most of the older double-barreled shotguns had a separate trigger for each barrel. Many of the newer double guns come with a single trigger that alternates between the two barrels, and for good reason. Experienced double-barreled shotgun shooters know that shooting the gun with fingers on both triggers at the same time will invariably result in both barrels firing together. The gun's recoil causes it to happen, and firing both barrels of a 12-gauge at the same time is a jolting experience. Older double-barreled shotguns with cracked wrists/pistol grip sections are a common sight. Experienced shooters will have only one finger in the trigger guard at a time.

The single-shot, break-action shotgun, although obviously not as capable for delivering quick follow-up shots as other configurations, does have several traits we might view as potentially advantageous. For one, they almost always cost less, either new or used, than any other type of shotgun. They are simpler and, we might say, not as exciting or as sophisticated as the other designs.

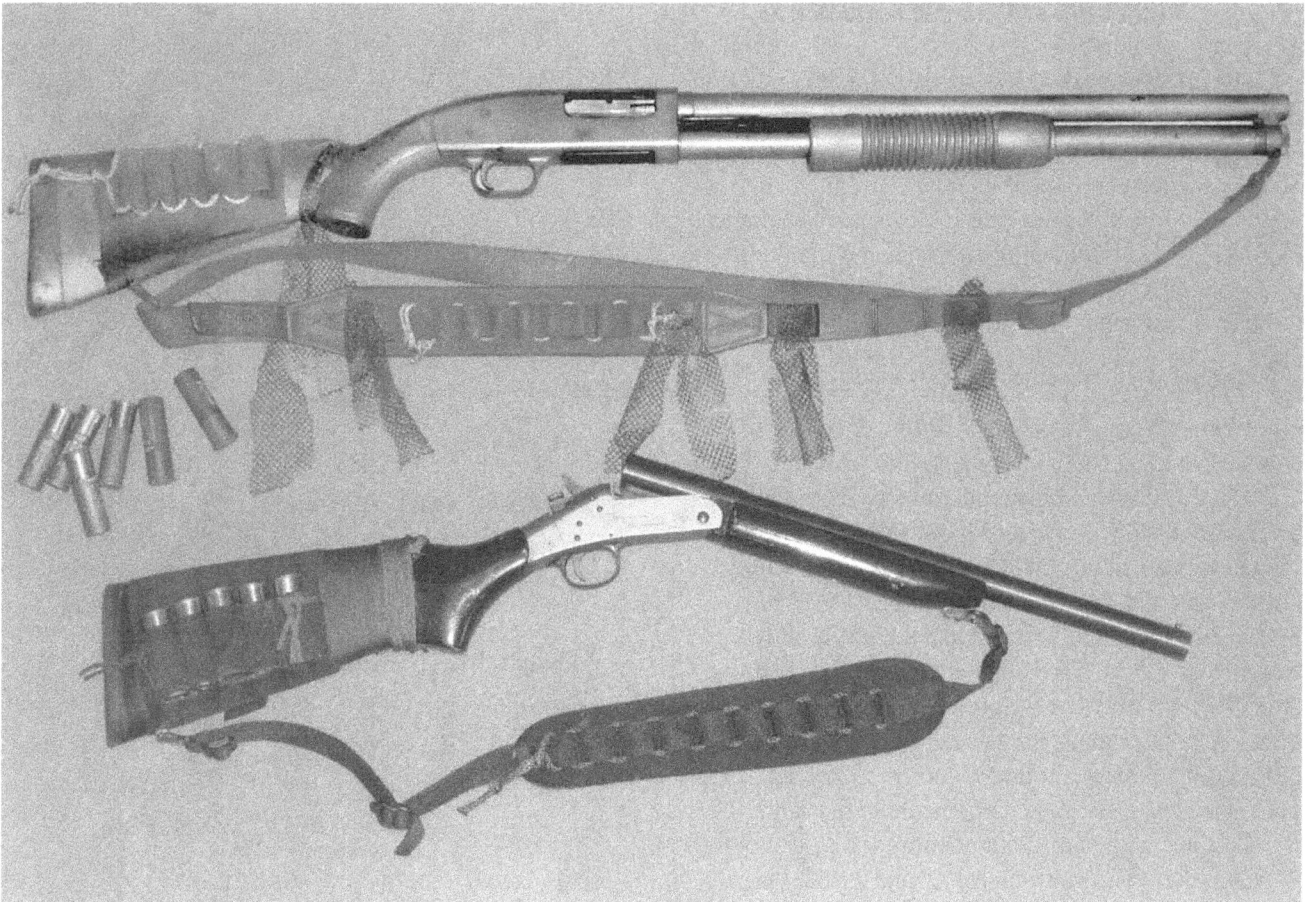

Two different shotgun designs are represented here—the pump-action at top and single-shot break-action at bottom.

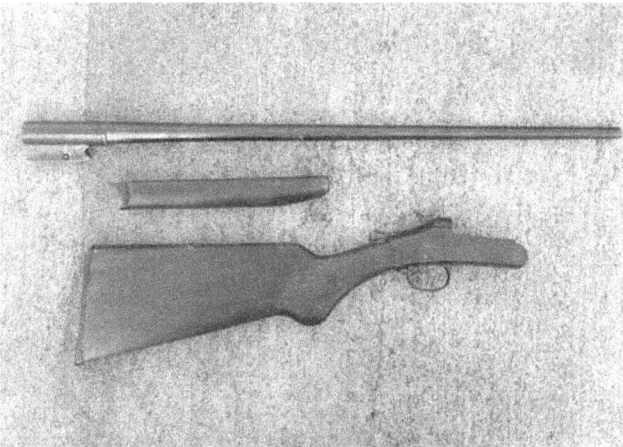

This single-shot .410 shotgun, like so many of its type, breaks down into three parts in about two seconds without using any tools at all, and this makes it easier to clean or stow in a pack.

Reliability is probably their best attribute. Because the single-shot guns are inherently simpler in design than the others, they have fewer parts to wear out or break, they are definitely easier to keep clean, and there simply isn't as much to them that can go wrong.

Every single-shot shotgun I have ever inspected seemed very lightweight compared to shotguns of other types. Again, there is simply not as much to them. But this weight factor can be a huge issue for someone traveling long distances on foot, as a postapocalypse survivor may well be forced to do just to find food or escape danger.

Most of the older model single-shot guns are easy to disassemble into three pieces, normally by merely pulling downward on the front part of the forearm piece—which is more often than not held in place by a spring clip—until the forearm pops free from the action. Once that much has been accomplished, the barrel will then unhook and separate from the stock and action section until you have three sections that are individually shorter than the assembled gun. Now, with access to more surface area, it's easier to clean the parts, and the disassembled gun will stow in a shorter space, like in a backpack. Reassembly is just as quick and easy.

THE TACTICAL SHOTGUN

The effectiveness of a short-barreled, pump-action shotgun in close-quarter gunfight situations was vividly depicted in the 1972 action film *The Getaway* starring Steve McQueen and Ali MacGraw. After the bank robbery he coordinates goes badly, McQueen's character, Doc McCoy, subsequently makes one too many mistakes that turn him into a fugitive. He decides he needs more firepower than that provided by his 1911 pistol alone, so he commandeers a shotgun from a Texas gun store. The scattergun with double-ought buckshot gives him the edge he needs—Hollywood style of course, but the basic concept is solid.

The forerunner to the modern tactical shotgun was a version of the Winchester Model 1897 pump-action 12-gauge with a 20-inch cylinder bore (fitted with an adapter that allowed attachment of the M1917 bayonet), which was used with buckshot loads in the trenches of World War I and hence acquired the name "trench gun." Whether we prefer calling these weapons trench guns, riot guns, police shotguns, security shotguns, home defense shotguns, combat, or tactical shotguns, their basic niche is fundamentally the same. They are shotguns with shorter barrels than hunting shotguns to make them more maneuverable and usable indoors, in alleyways, or in jungle terrain at close range. They have larger magazine capacities than the usual three rounds for hunting, because they are intended as tools that provide an edge in close-quarters *fighting*.

We have quite a variety of modern tactical shotgun configurations from which to choose these days. In the pump-action variety, some of the popular products include the Remington 870 Tactical and 887 Nitro Mag Tactical, Winchester's Defender (including the

Tactical pump shotguns: Mossberg 500 at left, Remington 870 on the right.

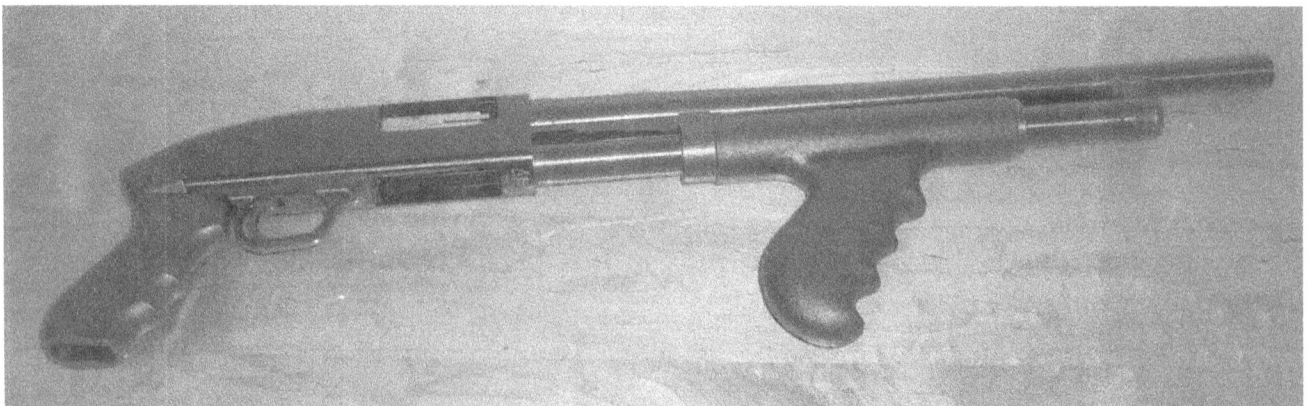

The Mossberg 500 12-gauge fitted with pistol grip handles to create a more compact fighting tool.

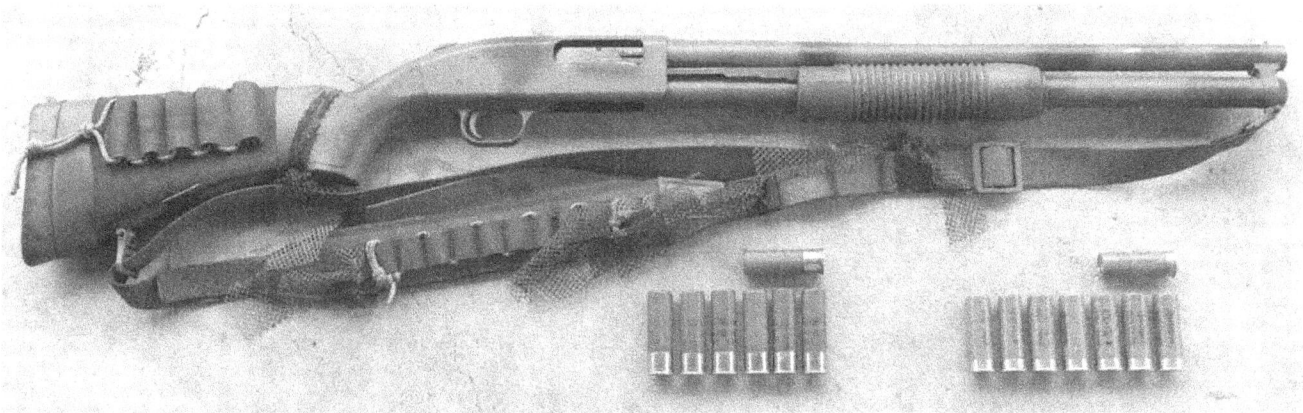

Author's Mossberg 500A with 20-inch barrel will hold 6+1 of 3-inch shells, or 7+1 of 2¾-inch shells.

White poster paper taped to a sheet of plywood works well for checking the shotgun's pattern at various distances.

most recent SXP Defender), Mossberg's numerous 500s and 590s (including the new Thunder Ranch Edition), Benelli's SuperNova Tactical, Stevens' 320, and Savage Arms' 350 Bottom-Eject Security Model, to name just a few.

The magazine tube of these tactical shotguns will typically hold at least five or six 12-gauge shot shells, and sometimes a few more depending upon shell and tube length. My Mossberg pump-action 12-gauge will hold six of the 3-inch shells plus one in the chamber, or seven 2¾-inch shells plus one in the chamber. There

are also extensions for shotgun tube magazines that can increase the available firepower.

Many hunting shotguns could hold more than three shells, too, if their hunting-regulation magazine tube plug were removed. In fact, a bird gun can easily be converted into a practical fighting shotgun by merely removing the plug in the magazine tube and shortening the barrel to 18 or 20 inches.

Being curious about the tactical shotgun's versatility for general multiscenario survival, I was particularly interested in finding out what the outer edge might be for reliable bird and small game hunting using birdshot. The weapon's short cylinder bore barrel definitely restricts its effective range, but I wanted to establish some reasonable expectation of the general limit in case I am someday forced to use this type of weapon for survival hunting.

My unscientific experiments with this seem to indicate that 30 yards is about the outer edge where the shot pattern, using #6 turkey loads through my 20-inch cylinder bore barrel, would still have better than about a 50 percent probability of bagging something the size of a squirrel, rabbit, grouse, or partridge. This is slightly better than the 25-yard practical limit I had expected, and I believe 25 to 30 yards is a reasonable range for bagging birds and small game under survival conditions. Interestingly, the same day I was out conducting this experiment, some other guys were trap shooting nearby with their 18-inch-barreled tactical shotguns, and I noticed they were hitting the clay pigeons with surprising regularity.

Within the tactical shotgun realm, autoloaders have become increasingly popular. It is certainly true that they are faster than other action types, but they also tend to cost more than their pump-action counterparts. The Remington 1100 Tactical semiauto, for

The Franchi SPAS-12 shotgun that can be used as either a pump or semiauto.

example, retails at over $800, which is almost twice the retail price of the average tactical pump shotgun. The Benelli M4 Tactical normally retails for $1,800 and up, depending on the variation.

By the way, Benelli's M3 Convertible shotgun can be changed between pump-action *and* semiauto. Another famous combat shotgun, the Italian-made Franchi SPAS-12 (Sporting Purpose Automatic Shotgun), also has the ability to switch between pump and semiauto mode. The SPAS-12 is no longer imported, unfortunately.

There is also the reliability question whenever considering autoloaders. Factory engineers have worked relentlessly with the automatics in recent decades, and consequently the reliability standard with the newer models seems to have reached a higher level than ever before. Most of the people I've talked with who use automatics for duck hunting seem very satisfied with the reliability of their guns.

Some of the newest tactical shotguns to appear recently carry features that I would characterize as exotic—very different from conventional styling. The new Kel Tec KSG Compact 12-gauge tactical pump shotgun, for example, looks almost like something from a science fiction movie and is advertised as being "as compact as legally possible" with a 26.1-inch overall length and an 18.5-inch cylinder bore barrel. (Check it out at keltecweapons.com.)

The KSG has dual tube side-by-side magazines below the barrel that hold six rounds each of 3-inch shells, or seven rounds each of 2¾-inch shells (6+6+1, or 7+7+1). The gun looks very much like a bullpup configuration, except that the magazines are forward of the action, whereas in a true bullpup the magazine would be positioned behind the gun's action. This shotgun weighs only 6.9 lbs. as advertised. It also features a pistol grip and Picatinny rail for mounting optional sights and things. The retail price usually runs less than $800.

Another perhaps even more impressive new tactical shotgun is the SRM 1216 from SRM Arms. This is a 16-round, semiauto 12-gauge with a detachable four-tube magazine unit that also serves as the forearm. Each of the four four-round magazine tubes is individually indexed into position for feeding, and once empty the next loaded tube is manually rotated into position. This weapon has an overall length of 32½ inches and an 18-inch barrel. It also comes with a Picatinny rail and a pistol grip configuration. The only thing keeping me from buying an SRM 1216 is the $2,400 price tag. Visit www.srmarms.com for more details.

EXOTIC SHOTGUN AMMO

As if the vast array of slugs, buckshot, and birdshot loads on the market today were not enough, we also have quite a selection of exotic, or what we might characterize as "novelty," ammo available for the shotgun.

Firequest (www.firequest.com) advertises a variety of interesting exotic ammo for both the 12-gauge and .410 on their website. For 12-gauge they offer flares, rock salt loads, blanks, and flechette rounds that contain clusters of finned metal darts, as well as a healthy variety of other goodies with names like Exploder, Armor Piercing, Zombie Killer, Double Slug, Macho Gaucho, Pepper Blast, Piranha, Rhodesian Jungle, Terminator, Pit Bull, and my favorite, Flame Thrower, also known as Dragon's Breath.

A friend of mine, Knut Rogers, and I recently had the opportunity to try out the Firestorm round in 12-gauge, which I imagine is something very similar (if not the same thing) as the famous Flame Thrower round. These novelty rounds come in three-round packs and, compared with conventional factory shotgun ammunition, are pricey per round (like $8 to $20 per three-round pack, depending on what you buy). Nevertheless, they are sometimes irresistible. Some of the novelty rounds apparently are not legal in certain states, so check before purchasing.

Two varieties of exotic ammo for the 12-gauge: Firestorm and Armor Piercing.

Shooting the 12-gauge Firestorm round.

The warning on the packaging for the Firestorm cautions that it is an "extreme fire hazard." We fired the rounds in early spring, while the surrounding woods were soggy wet with mud and some snow still on the ground, and experienced no problems whatsoever. I would certainly avoid firing this type of round anywhere near dry trees or brush, however. Our shooting was during daylight and the shot was not overly spectacular, but I imagine that during low-light conditions, something like this might light up the surroundings like a Fourth of July event. The maximum range of this type of round seemed to be no farther than about 30 or 40 yards.

Some of these novelty rounds might actually have a useful application in a survival situation. The Pit Bull, for example, is described as being loaded with six 00-buck pellets topped with a 1.3 oz slug. This would be somewhat similar to the old "buck and ball" loads that were often fired from smoothbore muskets, and it could be an impressive defense round.

MANAGING SHOTGUN RECOIL

Unquestionably, some of the hardest kicking shoulder firearms I have ever had the pleasure of shooting were 12-gauge shotguns loaded with slugs or heavy buckshot in 3-inch magnum shot shells. The reputation as a shoulder beater that the big-bore shotgun has long endured is, in my view, well earned. If you select a powerful 12-gauge as your apocalypse shotgun, you're going to need to learn how to manage recoil.

There are several approaches to consider for countering heavy recoil. Perhaps the first and easiest is to stick with standard 2¾-inch shot shells, which simply do not pound the shoulder as severely as do the longer 3- or 3½-inch shells. The shorter shells naturally do not offer quite as large a payload as the longer shells, but they will still deliver an impressive barrage of projectiles. Another advantage to them is that magazine tubes can hold more of them due to their shorter length.

The next line of approach is the butt stock. A thick, foam rubber recoil pad will soften a gun's felt recoil noticeably, certainly much better than an old-timer steel butt plate is able to do. Most modern shotguns feature some version of a soft rubber butt pad. Also, a wider pad will distribute the forces of the recoil over a larger area than will a narrower one, reducing the amount of kick felt at the shoulder.

The length of pull (distance from the trigger to the butt) can influence the perception of recoil to some de-

Shotgun shooting vest with kitchen hot pad sewn behind right shoulder to cushion recoil.

Demonstrating the makeshift hot pad shoulder cushion.

gree. Usually the longer the length of pull, the easier a given amount of recoil can be tolerated by the shooter, because it positions the gun's action farther from the shooter's face. Standard fixed shotgun stocks tend to have a comparatively long—often 14 inches or longer—length of pull, especially with the thick butt pads typically found on shotgun stocks.

Another popular method for reducing shotgun recoil is inserting a mechanical recoil reducer into the gun's stock or, with some products, attaching it to the magazine tube. This is a cylindrical-shaped device designed to counter the gun's recoil using inertia, and the two most common types use springs and pistons or mercury. Mercury-filled reducers tend to be quieter. For the stock-inserted reducers, a hole of the appropriate diameter must be bored into the butt of a shotgun that lacks an action screw hole.

The downside to any of these inertia or counterweight devices is that they add weight to your gun (most weigh between 6 and 10 ounces). Their added weight all by itself would in fact act to offset some percentage of the gun's recoil.

Recoil-reducing ported choke tubes combine a ported chamber—a variation of a muzzle brake—with the choke tube on the end of the barrel to simultaneously accomplish choking and recoil reduction. Examples include the Recoil Reducer Choke Tube from Royal Arms, the famous Lyman Cutts Compensator that is no longer in production, the Elite Ported Choke Tube from Colonial Arms (currently listed in the Brownells catalog), and similar products.

A shooter tends to feel less recoil whenever wearing a heavy coat or padded jacket while shooting high-powered long guns. The thick coat cushions the impact of the kick on the shoulder to some degree. It is not uncommon to see trap and skeet shooters wearing shooting jackets or vests that have padded areas around the armpit and inside shoulder where they place the butt of their shotguns. I discovered that a kitchen hot pad sewn to the underside of a vest in the right place serves quite well for this purpose.

CHAPTER 5

Multicaliber Guns for the Apocalypse

We have explored the three main categories of firearms—rifles, handguns, and shotguns—but this discussion would not be complete until we have seriously contemplated the versatile nature of multicaliber, or "combination" guns.

Wikipedia's definition of a combination gun is "a break-action hunting firearm that comprises at least two barrels, a rifle barrel and a shotgun barrel, but not always in an over and under configuration."

I prefer a broader definition besides just the multi-barreled long guns. For me, it includes any and all firearms having multiple caliber capability; in other words, the dual cartridge, or multicartridge, combo guns.

The whole concept of shoulder-fired combination guns more or less originated with the European (especially German) three-barreled drillings, the two-barreled shotgun/rifle cape guns, and the four-barreled vierlings, which all featured some combination of rifle with shotgun barrels within the same gun.

Over-under rifle/shotgun combos have been popular with survival-minded folks for decades, and the Model 24 series made by Savage Arms and even the somewhat less common, lightweight M6 Scout made by Springfield have been no exceptions.

Given their apparent demand and popularity in the area around where I live, it is surprising to me that these handy combination guns were ever dropped from production, but according to Savage Arms' website, the Model 24 was discontinued as of November 1, 2008.

An example of an old combination cape gun, its left barrel chambered for 16-gauge shotgun and its right rifle barrel in 11mm.

Used guns in good condition can still be found for sale from time to time, however. Also on the Savage website, I noticed that they show a new gun for 2012 that appears very similar to the Model 24, the Model 42 combination gun, which features .22 rimfire over .410 shotgun.

The Model 24 was a break-action, two-barreled, over-under combination gun, with the top barrel chambered for a rifle cartridge and the bottom, more often than not, a 20-gauge shotgun. Combinations of either .22 LR or .22 Magnum over 20 gauge are what I have seen the most, although I do remember seeing at least one example having a .30-30 barrel over a 12-gauge and more than one with a .222 Remington over 12-gauge.

I am only aware of two combinations that were

available with the Springfield M6: .22 LR over .410 shotgun, and .22 Hornet over .410 shotgun.

Perhaps the first true combination gun that I can think of was the Civil War-era LeMat percussion revolver, because it featured two separate barrels of different bore sizes. A unique and now famous weapon, the LeMat featured a nine-shot cylinder that rotated around a separate shotgun barrel. This gun was designed in 1856 by Dr. Jean Alexander Francois LeMat of New Orleans, but it was produced overseas and imported in limited quantities by the Confederates during the War Between the States. Early guns were manufactured in France and later on in England.

Original LeMat revolvers are very rare and valuable, but fortunately for us a replica of this fascinating gun is currently being manufactured by F.LLI Pietta in Italy and distributed for sale in America by such suppliers as Dixie Gun Works, Cabela's, and Navy Arms. The modern versions appear to be close representations of the originals, except that the originals featured .42-caliber cylinders and revolver barrels, whereas the new guns come in the more conventional .44 caliber. The cylinders rotate around a 20-gauge shotgun barrel, making this a unique combination gun. All the Pietta-made examples I have examined at gun shows seemed to have been of the highest quality.

The LeMat is a single-action, loose powder and ball percussion (rather than cartridge) revolver employing 150-year-old technology that some people may consider impractical in today's world of automatics and precision double-action cartridge revolvers, but I include it in our discussion because any handgun that can deliver nine rounds of .44 *plus* a round of 20-gauge before reloading should be a formidable sidearm in the apocalypse world.

Guns that will accept cartridge/chamber adapters and barrel inserts might also fall into this combination gun category. Numerous variations on this general concept have been marketed over the years, but at present the two main sources I have found are MCA Sports/Ace Bullet (www.mcace.com) and GaugeMate (www.gaugemate.com).

MCA Sports offers rifled chamber adapters that allow .32 ACP pistol rounds to be fired in a .30-30, .308, or .30-06 rifle, or .22 LR or .22 Magnum to be fired in a weapon chambered for .223 Remington, for just a few examples. MCA also sells shotgun inserts that allow smaller cartridges to be fired in break-action shotguns. The shotgun inserts are essentially barrel sleeves, and MCA Sports offers them in three lengths—

The modern-made LeMat revolver from Pietta in Italy. Image courtesy F.LLI Pietta.

This old single-shot .410 shotgun can be used with the 10-inch rifled barrel insert shown above to fire .22 rimfire ammunition. The offset firing pin that fits behind the cartridge and stays in position with a rubber seal ring allows the centerfire gun to work with the rimfire cartridge.

2¾, 10, and 18 inches—to fire a variety of cartridges (.22 Hornet, .32-20, .38 Special, 9mm Luger, .45 Long Colt, 7.62x39, .30-30, .410 shotgun, and even .45-70, to mention just some of the popular caliber options) in the 12-, 16-, and 20-gauge guns, and several .22 inserts for the .410 as well as the bigger shotguns.

The inserts chambered for rimfire cartridges require a centerfire plug. My concern with this tiny plug, which is hardly larger than an aspirin pill, is that it could be lost so easily, rendering the whole device useless. The fired empty rimfire case must be removed from the insert using a small-diameter rod, ideally of hardwood or aluminum, to push it out together with that centerfire plug. So this is by no means a fast and convenient way to fire and reload this shotgun using rimfire ammunition. Nevertheless, it does provide one more caliber option with the shotgun, and for the special-purpose for which it is intended, it is probably worth having around. The centerfire adapters naturally do not require the centerfire plug, so they will be more convenient to use in this sense.

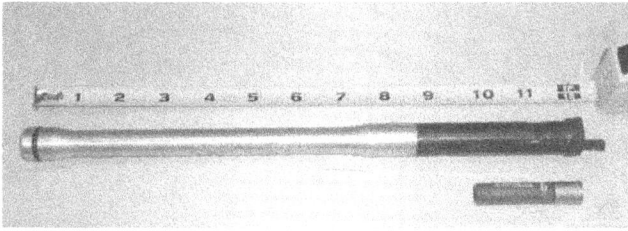

Savage's Four Tenner Model 412F sleeve allows .410 shells to be fired in a 12-gauge gun.

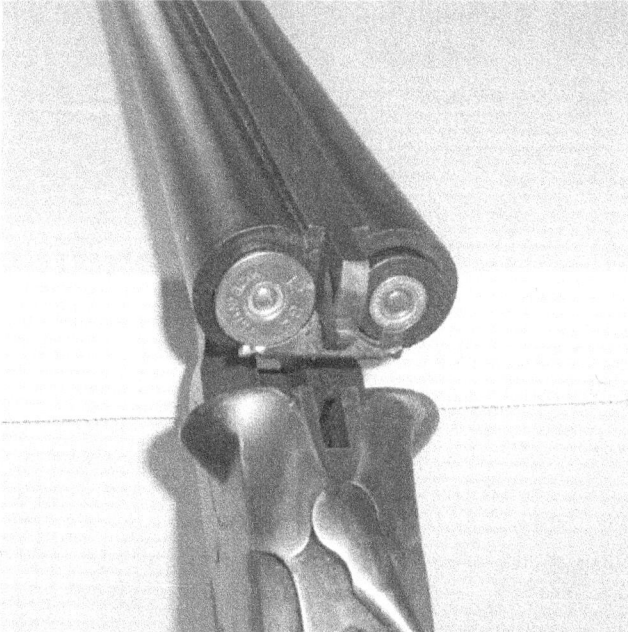

The double-barreled gun can be set up to fire either .410 or 12 gauge.

Ruger's Single Six single-action .22 convertible revolver. This gun came with two cylinders and will fire the entire family of current .22 rimfire cartridges, including .22 Mag.

GaugeMate (www.gaugemate.com) sells what it calls "sub-gauge adapters" that allow smaller-gauge shotgun shells to be fired in larger-gauge guns. (These have also been called caliber conversion sleeves, chamber inserts, sub-gauge inserts, and gauge reducers, among other things.) GaugeMate offers an all stainless steel adapter called GaugeMate Silver that is removed from the gun for reloading, and another product it calls GaugeMate Gold, designed to stay in the gun.

Savage made a useful product called the Four Tenner Model 412F, which is a 12-inch barrel conversion sleeve/insert that allows .410 shells to be fired in a 12-gauge gun. An apocalypse survivor with a double-barreled 12 gauge could keep something like this in one of the barrels and use the other barrel for the bigger cartridge so that he has both options ready to go all the time.

Revolvers having interchangeable cylinders chambered for different cartridges provide this combo-gun versatility as well. About 30 years ago I acquired a Ruger Blackhawk single-action revolver in .357 Magnum that included another cylinder chambered in 9mm Luger. The two bullets, being very close in diameter, can share the same barrel in a revolver designed as such. Although I fired hundreds of rounds of .357 and 38 .Special through that particular gun, I don't recall ever actually using the 9mm cylinder. Even so, I always liked having that other possibility available.

Probably one of the most popular dual-cylinder convertible revolvers has been the Ruger Single Six, commonly provided with one cylinder chambered for .22 Long Rifle (which means that it will also chamber .22 Short and .22 Long) and the extra cylinder chambered for .22 Magnum. This dual-cartridge convertible feature is only practical with revolvers having easily removable cylinders that can be conveniently switched out. Unfortunately, this prevents it from being a common feature with modern double-action revolvers with conventional swing-out cylinders, which don't lend themselves well to quick and easy cylinder changes.

Another interesting innovation that appeared several years ago is the revolver that will chamber either a standard handgun cartridge or a shotgun cartridge in the same cylinder, specifically the .45 Long Colt and .410 shotgun combination chamber. To date, several products have appeared on the market having this feature, but Taurus International (www.taurususa.com) has certainly led the way with its famous Judge series of revolvers. These double-action, five-shot revolvers are easily recognized by their longer than conventional cylinders that will accommodate the lengthy shot

The T/C Arms Encore three-barrel system allows a shooter to have a cartridge rifle, shotgun, and muzzle-loader all in one gun. Photo courtesy T/C Arms.

The Judge from Taurus will fire either .45 Long Colt or .410 shotgun rounds, or some mix of both, in the same five-shot cylinder.

shells. Some variations will chamber 2½-inch .410 shells, while others will chamber 3-inch shells. Taurus has recently added the Raging Judge Magnum that chambers not only the .45 LC and .410 2½- and 3-inch shells, but also the .454 Casull. Recently, even a revolving carbine version called the Circuit Judge has been offered under their Rossi brand of products. An innovative protective shield on the Circuit Judge covers the front sides of the cylinder, preventing the usual spray of hot gases and bullet fragments from the front of the cylinder from impacting the shooter's supporting forearm, as was common in the past with various revolving rifle designs.

This concept of having a fast and compact double-action revolver capable of delivering shotgun performance naturally appeals to a lot of people who want a versatile and effective close-range weapon for self-defense. The way this shotgun/handgun complies with the federal law restricting shotgun barrel length is due to its rifled barrel. Even though it will fire shotgun ammunition, it simply doesn't fit the government's technical definition of a smoothbore shotgun, fortunately for us.

There are a few single-shot rifles, pistols, and shotguns that become instant combo guns with a quick change of their barrels. Thompson/Center Contender pistols and rifles, for example, have provided this interchangeable barrel feature for decades, allowing shooters the ability to shoot more than 20 different cartridges through the same break-open, single-shot pistol or rifle.

Anyone seeking maximum versatility with a single firearm might just find it with the Encore system from Thompson/Center Arms. It is described as the "most complete shooting system available." The platform offers more than 90 barrel selections in over 17 cartridges, from .22 Hornet to .416 Rigby, 12- and 20-gauge shotgun smoothbore or rifled barrels, and even .50-caliber muzzleloader barrels. The popular three-barrel system allows the shooter to have a cartridge rifle, shotgun, or muzzleloader by simply changing the barrel. The company pitches the system as "A Complete Shooting System for Serious Sportsmen" and for "Any Game Anywhere." I noticed that T/C Arms features a modernized flintlock muzzleloader on its website as well (www.tcarms.com).

For an example of a somewhat similar system with a shoulder-fired weapon, Rossi USA (www.rossiusa.com) offers several single-shot models with two, three, and even four interchangeable barrels that all fit the same action. They offer one called the Trifecta that comes with three barrels for 20 gauge, .22 LR, and .44 Magnum or .243 Winchester, and they offer another called Pick 4 that will fire .243 Winchester, .22 LR, 20 gauge, and .410 shotgun. Talk about versatility!

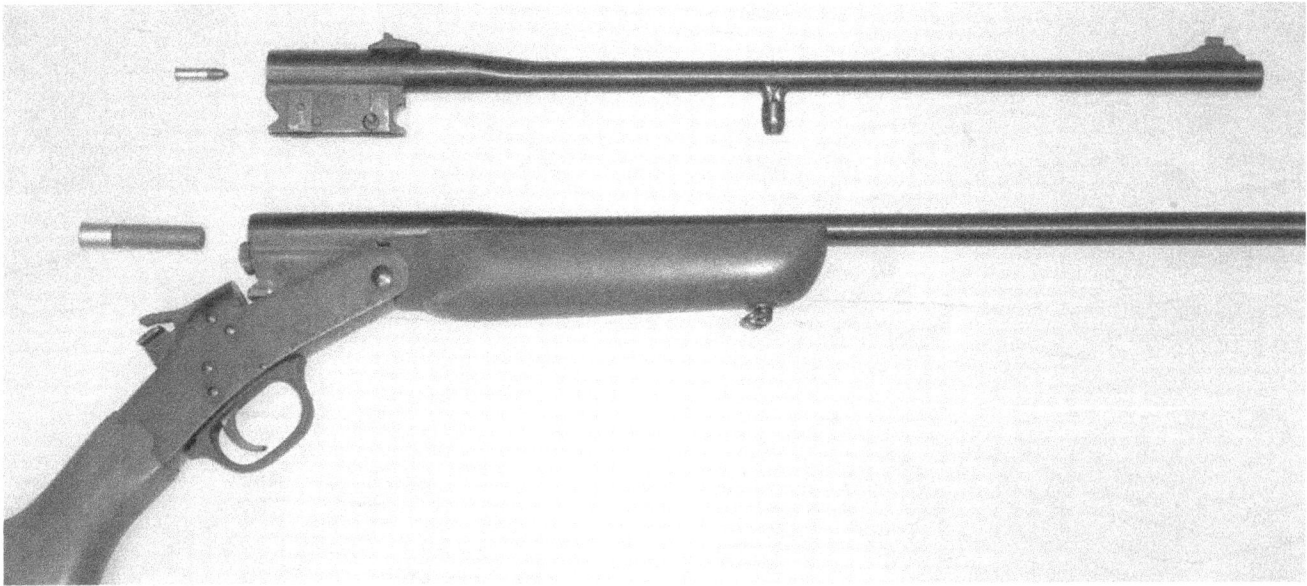

This Rossi break-action gun came with two barrels: one chambered for .410 shotgun and the other for 22 LR.

Long guns with multiple barrels that chamber different cartridges, revolvers with interchangeable cylinders that chamber different cartridges, guns with interchangeable barrels, and shotgun revolvers are perhaps the combination-type guns that first come to mind whenever we consider multicartridge versatility. However, certain firearms will simply chamber more than one type of cartridge in their regular chambers, without changing anything but the ammo. The common .357 Magnum is one example of this, because it will also fire .38 Special ammunition in its chamber, as we have already discussed. I have prepared a short list of some of the popular chamber sizes that will accommodate more than one type of cartridge, offering a degree of versatility.

It is worth keeping in mind that this combination gun/multicaliber capability is not only versatile in its possible applications, such as the wide range of game sizes that can be hunted or the many different defense or tactical situations that can be addressed, but it is also more adaptable in its feeding requirements. In other words, having the ability to use more than one type of ammunition could mean the difference between having something to shoot with or having no usable firearm at all, depending on the availability of ammunition for which the weapon is chambered. The little country store down the road may stock only a small assortment of ammunition. Having a dual-caliber weapon means having twice as much chance of being able to find usable ammunition as you would otherwise have with a single-caliber weapon.

FIREARM CHAMBERINGS WITH VERSATILITY

Chamber Size	Cartridges That Will Chamber
.22 Long Rifle	.22 CB, .22 BB Cap, .22 Short, .22 Long, .22 LR
.22 Winchester Magnum	.22 Winchester Rimfire (WRF), .22 Winchester Magnum Rimfire (.22 Mag or .22 WMR)
.327 Federal Magnum	.32 S&W, .32 S&W Long, .32 H&R Magnum, .327 Fed Mag
.357 Remington Maximum	.38 Short Colt, .38 Long Colt, .38 Special, .357 Magnum, .357 Maximum
.44 Remington Magnum	.44 Russian, .44 Special, .44 Rem Mag
.460 Smith & Wesson	.45 Schofield/.45 S&W, .45 Long Colt, .454 Casull, .460 Smith & Wesson

CHAPTER 6

Ammunition for the Apocalypse

Now that we've discussed a range of weapon types and briefly highlighted the differences between certain types of ammunition along the way, let's talk more specifically about which calibers and specific cartridges and loads might be preferable for our needs in the apocalypse.

Firearms enthusiasts all have their own favorite cartridges for infinitely different reasons. Performance wise, we could make a strong case for almost any popular modern cartridge within its own niche (besides maybe the .25 ACP, which arguably fills no role at all particularly well), be it one specific shotgun shell or another, a certain pistol cartridge, or any of the multitudes of rifle rounds that have ever been in common use.

However, if we're going to arm ourselves for an apocalypse, we should consider more than just cartridge performance alone. It is true that muzzle velocity and bullet weight, bullet diameter and shape, jacketed or cast lead design, kinetic energy, and bullet trajectory are all important factors in this. But as we prepare for global cataclysms, we should perhaps also consider such issues as availability of supply both before and long after the world event, and the size and weight of the ammunition—factors that will influence the portability issue, the practicality of reloading under potentially makeshift conditions, and the cost of stocking up, especially in bulk supplies.

Surplus military ammunition is often desirable for some of these very reasons, as we've already considered. Let's talk about some of the more popular examples.

CARTRIDGES TO CONSIDER

The focus on certain cartridges in this chapter is not by any means suggested as a complete inventory of viable apocalypse ordinance. Rather, it is to draw

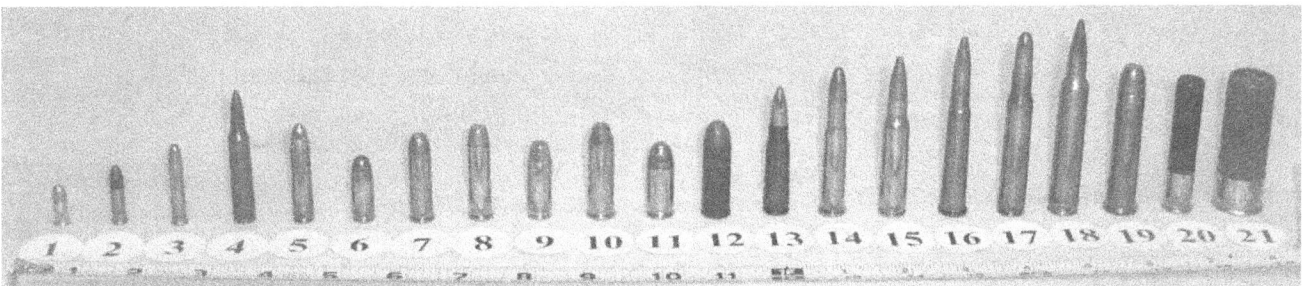

Popular cartridges for size comparison: (1) .22 Short, (2) .22 Long Rifle, (3) .22 Magnum, (4) .223 Rem/5.56mm NATO, (5) .30 Carbine, (6) 9mm Luger, (7) .38 Special, (8) .357 Magnum, (9) 10mm Auto, (10) .44 Magnum, (11) .45 ACP, (12) .45 Long Colt, (13) 7.62x39 Russian, (14) .30-30 Win, (15) .308 Win/7.62x51mm NATO, (16) .303 British, (17) .30-06 Springfield, (18) .300 Win Mag, (19) .45-70 Government, (20) .410 shotgun, (21) 12-gauge shotgun.

attention to some examples that are among the most popular cartridges with preppers and note why they are so popular.

.45 ACP

We've already discussed the 9mm pistol cartridge at some length, but the .45 Auto, also known as .45 ACP (Automatic Colt Pistol), was the official primary pistol cartridge of the United States armed forces from before World War I until it was replaced by the 9mm in the 1980s. The .45 Auto was the standard handgun round of U.S. forces fighting in both the world wars and the Korean and Vietnam Wars.

This happens to be an important round to consider for our purposes for several reasons. For one thing, it is considered by many to be a formidable man-stopper within its effective range (up to 50 meters/54 yards, according to the Army), mainly owing to its comparatively large diameter by contemporary handgun standards. The .45-caliber bullet need not expand as quickly or as much as a smaller-diameter bullet upon impact in order to achieve the same or similar result.

Also, the fact that there are so many different guns chambered for the .45 ACP—including numerous 1911-style and plenty of other auto pistol designs, certain double-action revolvers that use the cartridge in conjunction with moon or half-moon clips, submachine guns, and several carbines—makes it an attractive general-purpose survival round.

The case length of the .45 ACP is shorter than that of the older .45 Long Colt revolver cartridge, and its usual bullet weight range is slightly lighter (the ACP's 185 to 230 grains compared with the Long Colt's 225 to 255 grains). It is also a rimless auto cartridge, whereas the .45

The .45 ACP round (on the right) next to a 9mm Luger round for a size comparison.

LC has a rimmed case. The most common bullet weight for the .45 ACP is 230 grains, and the muzzle velocity typically ranges from around 850 to 1,000 fps in handguns.

7.62x39mm

Often loosely referred to as "SKS ammo" or "AK-47 ammo" (and I've even heard it referred to as the 308 Russian, although doing so may lead to confusing it with the now modestly popular 7.62x54R), the Soviet 7.62x39mm rifle cartridge has become incredibly popular in the West within just the past 20-some years. This is the result, in large part, of the huge quantities of surplus Chinese and former Eastern Bloc firearms that were imported into the United States before the Clinton-era bans on their importation went into effect. And the usual low price of surplus 7.62x39 ammunition compared against a lot of other military surplus rifle ammunition is clearly another factor in its current popularity in the United States.

The 7.62x39 is generally considered a very mediocre high-powered rifle cartridge when compared with the battle rifle cartridges of most nations in the same or close bullet diameter class, before the smaller caliber trend that evolved with the M16 rifle. Available factory loads do vary, of course, but 122- to 124-grain bullets at slightly less than 2,400 fps muzzle velocity seem pretty close to the norm for this round. This doesn't even quite measure up to the same level as the modest .30-30 Winchester cartridge with any of its common modern factory loads.

Nevertheless, literally tens of millions of shoulder weapons have been chambered for the 7.62x39 worldwide, and quantities of military surplus ammunition are abundant and, comparatively speaking, cheap. Additionally, soft-point hunting loads in 7.62x39 are available from the major ammunition manufacturers now, enhancing this cartridge's credibility for sporting use.

.30-06 Springfield

The "ought six" (metrically known as 7.62x63mm) is now aging but still going strong by every measure. Although it hasn't been a standard-issue American service rifle round since the 1950s, it has long been considered by many as *the* original standard by which all modern high-powered rifle cartridges are compared, perhaps even to this day.

In these days of short magnums and hypervelocity rifle bullets, the .30-06 might seem rather boring, and it is definitely among the original old-school cartridges, but it is still every bit as versatile and reliable

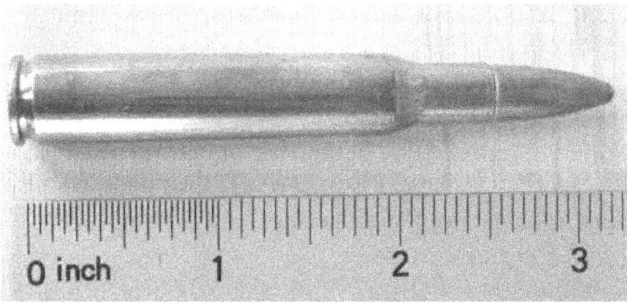

A 30-06 cartridge next to a ruler to show overall length.

a performer as it ever was. I include it in this discussion of surplus military cartridges because of its origination and extensive military use in the past, but in reality this is still one of the most popular sporting big-game rifle cartridges in North America.

The case of the .30-06 was one of the early rimless designs, making it more suitable for feeding from a box magazine than were any of the previously more common rimmed cases.

One of the most desirable characteristics of the old .30-06 is its flexibility with different load possibilities. We briefly discussed some of the ballistics common with this cartridge in chapter 3, but 150-grain bullets having a muzzle velocity slightly over 2,900 fps to 180-grain bullets with muzzle velocity of 2,700 fps seem to span the most popular range, with muzzle energies of 2,700 to 2,900 ft-lbs. being common with the ought six.

.308 Winchester/7.62x51mm NATO

We briefly discussed this cartridge already, but now let's take a closer look at its specifications. The 308 Winchester case, with its overall length of 2.015 inches, is .479 inch (almost a half-inch) shorter than that of the .30-06. The .308 is normally considered a functional equivalent of the old .30-06 for all practical purposes, although if we compare apples-to-apples with bullet weight and corresponding powder loads, the 308's bullet will usually exit the muzzle around 100 to 150 fps slower than the .30-06 bullet, all else like barrel length and action type being equal.

Popular loads for the .308 include 150-grain bullets at 2,800 fps muzzle velocity, 165-grain bullets from around 2,600 to 2,700 fps, and 180-grain bullets at around 2,500 fps. It is described in *Cartridges of The World* as having a reputation for excellent accuracy and a favorite of target shooters, as well as being suitable for hunting most North American big game. There is no doubt that the .308 Winchester is one of our most popular modern .30-caliber high-powered rifle cartridges.

.303 British and 7.62x54R Russian

These two cartridges have enough in common that I often think of them in the same general category. Both have rimmed cases (unlike most of the world's military cartridges of the twentieth century), both were extensively used in both world wars, both

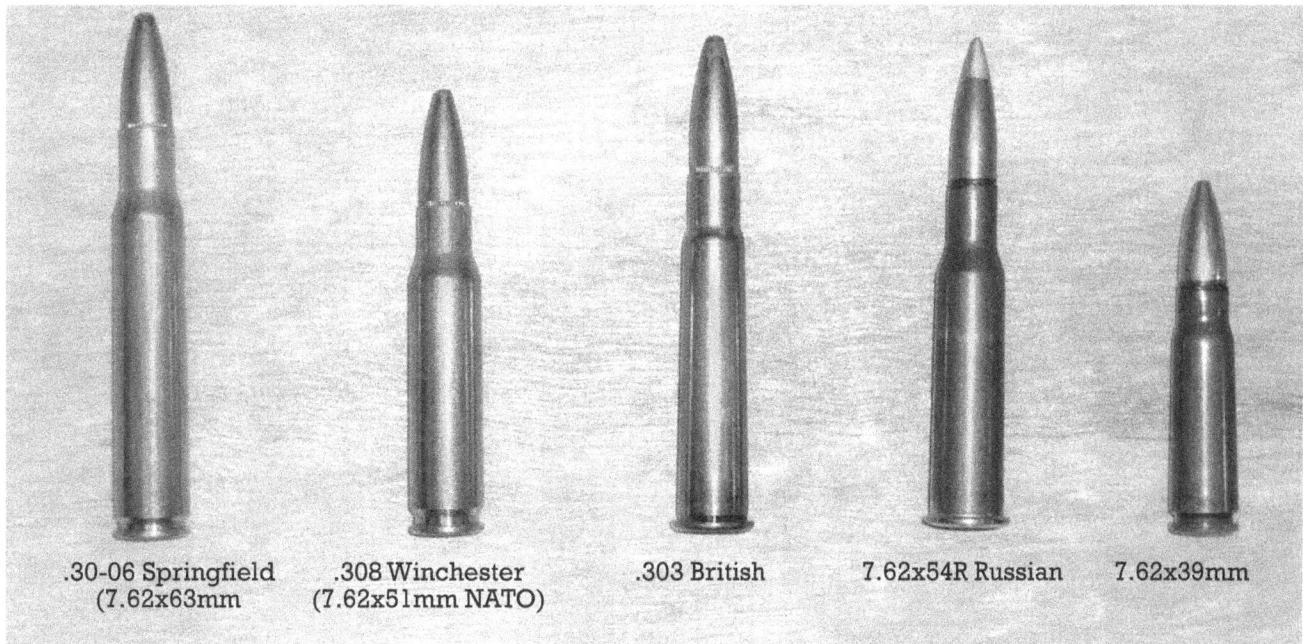

.30-06 Springfield (7.62x63mm)   .308 Winchester (7.62x51mm NATO)   .303 British   7.62x54R Russian   7.62x39mm

Five fairly common .30-caliber military rifle cartridges of the world shown here for relative size comparison.

are of foreign origin, and they are roughly comparable in their ballistics.

In terms of popularity and standardization, the .303 British was/is to the British Empire and its commonwealth what the .30-06 is to the United States.

Winchester currently offers hunting loads for both the 7.62x54R and the .303 British. The Russian cartridge has more than a 150 fps advantage and more than 300 ft-lbs. of energy over the British cartridge, with the 180-grain hunting bullet of both rounds in the Winchester catalog.

.30 Carbine

The M1 Carbine was issued to U.S. service personnel (especially to platoon leaders in the field), support units, and medics during World War II and later. The weapon was initially intended to fill a perceived gap between the .45-caliber pistol and the M1 Garand infantry rifle.

Being compact and lightweight (the weapon weighs only about five pounds without a loaded magazine), semiautomatic (a full-auto version designated the M2 Carbine was also used), and relatively simple to operate, as well as having low recoil, it enjoyed a degree of popularity in its era, in spite of the . 30 Carbine's notorious lack of stopping power.

Original war-era carbines have become increasingly collectible in recent years (and consequently more valuable), but there are still quite a few newer versions of the M1 Carbine made for the civilian market. Put out by companies like Universal and Plainfield, prices are reasonable by today's standards. Additionally, there have been auto pistols, such as those made by AMT, chambered for .30 Carbine. So there are still

plenty of guns in existence chambered for the little .30, and it never has become totally obsolete.

The cartridge is small by rifle standards and, as already noted, notorious for lacking the power expected of an effective battle rifle or carbine. Presently, Winchester lists two loads for the .30 Carbine in its catalog, both having 110-grain bullets at 1,990 fps muzzle velocity.

Sources of *inexpensive* surplus .30 Carbine ammo, on the other hand, have become very hard to find these days. Hence, the cartridge has apparently fallen dramatically in popularity within the past several decades.

A .30 Carbine cartridge (right) next to a .32-20 for a relative size comparison.

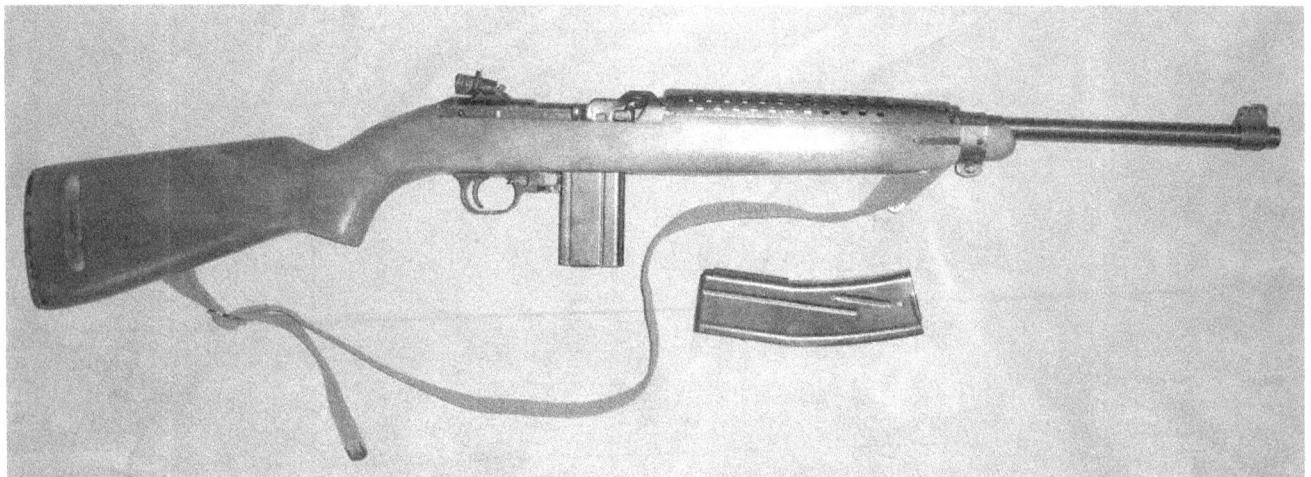

An M1 Carbine made by Universal for the civilian market, shown here with 15-round detachable box magazine in the weapon, plus an auxiliary 30-round magazine.

5.56mm (NATO)/.223 Remington

The military surplus cartridge we will talk about now is the prolific (and equally controversial) "two-two three," commonly referred to by military personnel as "five-five-six." Here we depart from the more traditional .30-caliber military rifle rounds and move into a more contemporary realm of smaller, lighter, more *efficient* military rounds.

The 5.56x45mm NATO has been with us now for more than 50 years, and it has served our armed forces as of this writing for over 45 years. Viewed by some old-timers as hardly more than a varmint cartridge, the .223 has nevertheless proven itself capable as a combat round, at least within its effective range, which seems to be about 300 meters/328 yards.

The Vietnam-era military round was the M193 Ball, which featured a 55-grain, full-metal-jacketed, spire-pointed bullet with a muzzle velocity of 3,250 fps from an M16A1 rifle barrel. This load was effectively replaced for military use in the 1980s by the NATO SS-109 and the US M855, each having bullet weights of more than 60 grains, in conjunction with the then-new M16A2 rifle variant that featured rifling with a faster rate of twist more suitable for stabilizing the heavier bullets. The longer, heavier bullets provide higher ballistic coefficients and enhance the performance of the service weapons at extended ranges. Commercial loadings for .223 Remington are currently available with bullet weights normally ranging from 35 to 77 grains, at muzzle velocities from 2,500 fps with the 77-grain to 3,800 fps for 35-grain bullets.

It's important to know that cartridge, chamber, and throat dimensions of .223 Remington sporting rifles and 5.56x45mm NATO military rifles are not identical, which is why you'll encounter chamber pressure warnings about using NATO rounds in sporting rifles chambered for .223 Remington.

The .22 Rimfire Cartridges

In the first chapter, I suggested that the self-defense requirement we will surely want our doomsday weapons to address almost forces us to dismiss the small-caliber firearms in our search for a first-grab, general-purpose type of weapon. However, despite the obvious limitations of the small rimfire cartridges, they are also comparatively compact, economical, efficient, and impressive performers for their size. Also, they generate less noise and a lot less recoil than do any of the larger centerfire cartridges. For these reasons, the .22s are wildly popular survival firearms

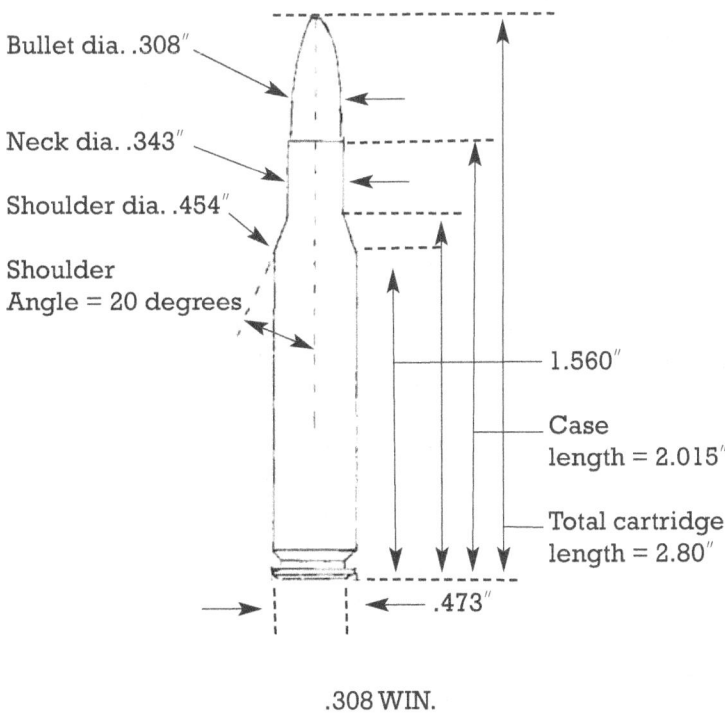

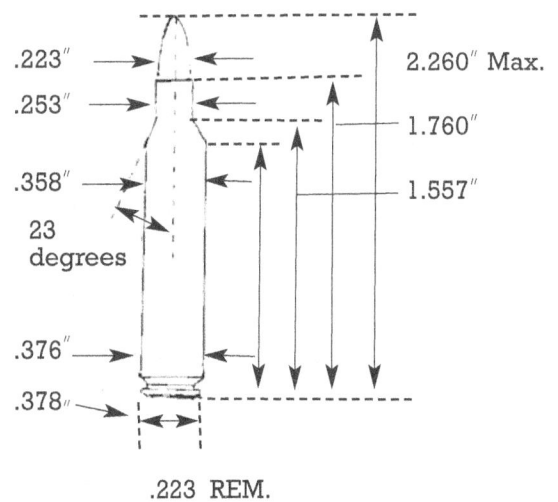

Bullet dia. .308″

Neck dia. .343″

Shoulder dia. .454″

Shoulder Angle = 20 degrees

1.560″

Case length = 2.015″

Total cartridge length = 2.80″

.473″

.308 WIN.

.223″

.253″

.358″

23 degrees

.376″

.378″

2.260″ Max.

1.760″

1.557″

.223 REM.

Dimensions of the .308 and .223 cartridge for size comparison.

An assortment of firearms chambered for .22 LR are represented here, from the revolver and auto pistols to rifles of all styles.

within their particular niche, and virtually every style of firearm can be found chambered in .22 LR.

Any gun chamber that accepts .22 Long Rifle ammunition will also accept the .22 Short, the .22 Long, and the low-velocity .22 CB and BB Caps. However, not all actions of .22 LR guns will necessarily cycle properly with any of the less powerful and shorter .22-caliber cartridges. But at least in manually operated single-shot mode, this flexibility of .22 LR-chambered guns will be potentially very useful.

If there is any doubt about the efficiency of the little .22, one need only read the warning on just about every box of .22 LR ammunition, which states that the bullet is dangerous up to a mile and a half away! I can't even see a person that far away with my naked eyes.

Also, in recent decades we have seen an impressive assortment of high and ultra-high velocity, high-performance versions of the .22 LR introduced from several manufacturers, including the famous Mini-Mags, Stingers, and Velocitors from CCI, Winchester's Hyper Speed and Varmint HE, and Remington's Vipers and Yellow Jackets, any of which could be relied upon to enhance the performance of the common rimfire cartridge. With these loads, the bullet weights range anywhere from around 30 to 45 grains, and reported muzzle velocities range from over 1,200 to nearly 1,500 fps, depending on a number of variables.

For those who see the practicality of small rimfire cartridges but prefer a tad more power than what even the high-performance LR rounds provide, the .22 WMR (.22 Winchester Magnum Rimfire, or simply .22 Mag) fills the niche very well. Magnum rimfire ammunition is more expensive than the more popular .22 LR, but its performance is superior. Most of the .22 WMR

From left to right: .22 Short, .22 Long Rifle, and .22 Magnum. All three are truly impressive rounds for their size.

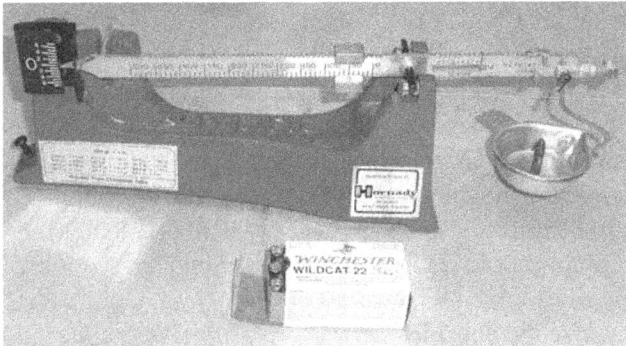

Weighing a .22 Long Rifle cartridge with the hand-loading scale. This one weighs 51 grains.

Federal .410 2 ½ in.

Hi-Brass with ½ oz. #6 lead shot

Weight of loaded cartridge: 322 gr.

Winchester "Wildcat 22"

High-velocity .22 Long Rifle 40-gr. lead round-nose bullet

Weight of loaded cartridge: 51 gr.

Interesting weight comparison between a .410 shotgun cartridge and the .22 LR cartridge. Both are well suited for small game hunting, but you can have six rimfire rounds for the weight of a single shot shell.

loads I've seen listed in catalogs show bullets in the 30- to 40-grain range exiting rifle barrels at around 2,000 fps, with usually something more than 300 ft-lbs. of muzzle energy.

The .22 Magnum case is visibly longer than the .22 LR, but it also has a slightly larger diameter, so a .22 LR will not safely fire in a magnum chamber.

.30-30 Winchester

Originally known as .30 WCF (Winchester Center Fire), the .30-30 has been in existence since 1895, when it was first chambered in Winchester's Model 1894 lever-action rifle. It was one of the first few smokeless powder cartridges to appear on the market in America. Compared against the majority of rifle cartridges common in its day that typically had bullet diameters ranging from 38 to 45 caliber and cases loaded with black powder, the new .30 smokeless cartridge was thought of as a remarkably flat-shooting round with impressive high velocity. But of course that was then, and a lot has changed with rifles and ammunition over the ensuing 117 years.

Nowadays the .30-30 is generally perceived as a very mediocre big-game hunting rifle cartridge, perhaps best suited for deer hunting in dense forests,

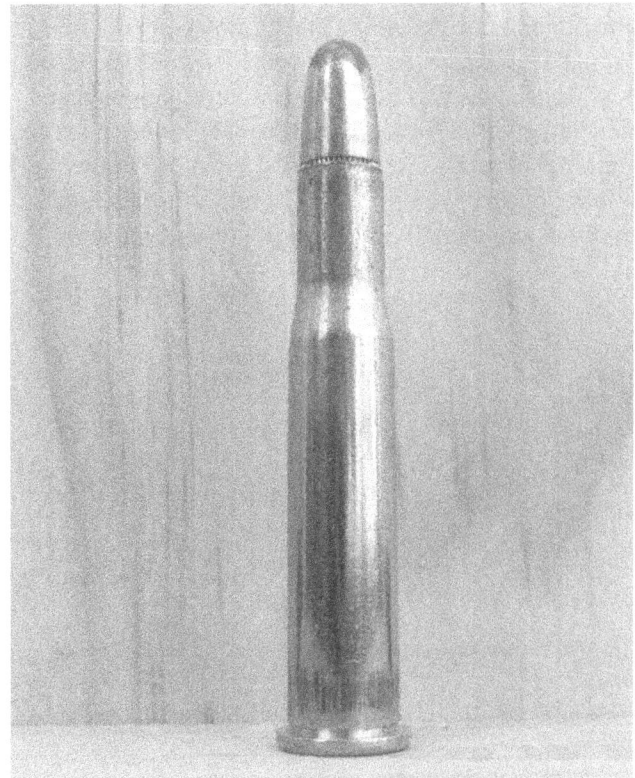

The .30-30 Winchester cartridge.

where the shots will rarely stretch beyond 150 yards. Factory loads are most commonly available with either 150- or 170-grain bullets. As a general rule, the lighter 150-grain bullet would be better suited for hunting the smallest of the big-game animals, like whitetail deer or antelope, whereas the 170-grain would be the more logical choice (of the .30-30 bullet weights, anyway) for the heaviest mule deer bucks, elk, and black bear.

The .30-30 is a rimmed cartridge, so it will more suitably feed from its most common tubular magazine than from a box magazine that stacks the cartridges one over another.

Probably the 30-30's most important attribute is its enduring popularity. A survivor is likely to find a source of .30-30 ammunition supplies in any little town or country store just about everywhere in North or South America, and this issue of availability could be a huge one in a crisis.

.40 Smith & Wesson

Much was said about the 10mm Auto in the chapter on handguns, and it happens to have been the first of the contemporary auto pistol cartridges to fill the .40-caliber niche when it was developed in the early 1980s.

But when the FBI began working with the 10mm cartridge, it decided that a loaded-down version would be adequate for its purposes. It would produce less recoil, which is better in combat situations that demand quick follow-up shots, as well as be easier for training purposes. So it was only a matter of time until a less-powerful version of the 10mm was developed, and this was the .40 Smith & Wesson that made its

Pistol rounds for size comparison, left to right: 9mm, .40 S&W, 10mm, and .45 ACP.

debut in 1990. The .40 S&W has a shorter case length than the 10mm but uses the same diameter bullet.

If there has been any other firearm cartridge since the birth of the .40 S&W that has risen in popularity as quickly or to the same level, I am certainly not aware of it. I don't think it is an overstatement to say that the .40 is a phenomenally popular pistol cartridge right now. The vast majority of my shooting buddies own and favor this cartridge, and its recent popularity with law-enforcement agencies cannot be ignored.

The .40 seems to fill that caliber gap between the 9mm and the .45 ACP perfectly. While the muzzle energy numbers range in the mid to high 300s with most of the current 9mm factory loads, the majority of the .40 loads range between 400 and 500 ft-lbs. The .40's muzzle velocities typically run more than 100 fps faster than the 45 Auto's average, and its bore diameter is only about 1/20th of an inch (.05 inch) smaller. Bullet weights for the .40 S&W usually hover around 165 to about 180 grains—lighter than the .45 but heavier than the 9mm. And .40-caliber handguns generally hold a lot of rounds (as opposed to a .45) of potential man-stopping power (as opposed to a 9mm), making it the ideal compromise round in many people's eyes.

.44 Remington Magnum

Much was discussed concerning the incredibly versatile .357 Magnum in chapter 2, and that was the original handgun magnum. But when the heavier and more powerful .44 Magnum arrived 20 years after the birth of the .357, the world of handgun hunting took a significant leap forward.

We did consider some aspects of the .44 in the handgun chapter, and it *is* one of my favorite firearm cartridges, for several reasons. For one thing, it is a very capable cartridge that, due to the typically robust nature of the guns chambered for it, can be safely loaded to its full potential. The same is not always true of older big-bore revolver cartridges like the .45 Long Colt, whose original guns were not built for the hottest smokeless powder loads of today. And the .44's size and velocity norms make it a respectable big-game brush hunting round.

Second, its long-running popularity and consequently the wide variety of firearms chambered for it make the .44 Magnum a viable general-purpose survival cartridge. We find numerous handguns, rifles, and carbines chambered in .44 Magnum these days, as well as a healthy variety of available factory loads.

I have experimented with a number of factory loads as well as my own hand loads in my .44s over the past 20 or so years. Semijacketed as well as hard cast lead semiwadcutter "Keith Style" bullets in the 240- to 250-grain weight range get the most mileage in this caliber. For all of my .44 Magnums, the 240-grain semijacketed soft-nosed bullet with a muzzle velocity of 1,250 fps (from a revolver barrel, and possibly 1,400 fps or faster from a carbine) gives me peace of mind while scouting about in the mountains of Idaho, where an encounter with an aggressive bear is not all that unlikely.

STOCKING UP ON AMMUNITION FOR THE APOCALYPSE

Ammunition of all kinds has never been especially cheap, and since 2008, prices have gotten almost out of control. There are several unrelated reasons for the high cost of ammunition, but the bottom line for us is that it is a reality we will have to face one way or another, because we will need ammunition to feed our weapons in the apocalypse.

Some shooters have been able to offset their shooting costs by hand loading their ammunition. A

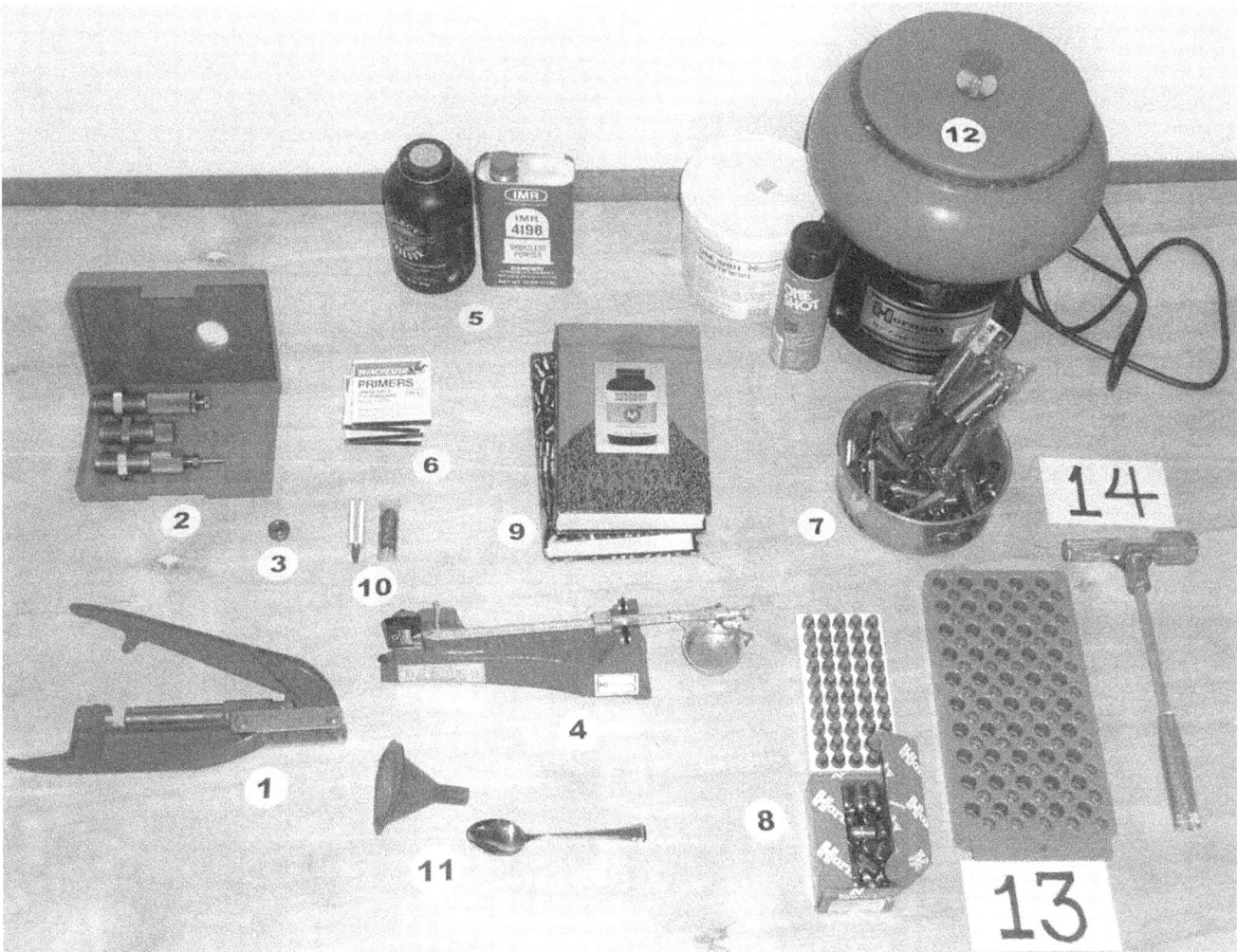

Everything needed to hand load for one caliber (in this case, .45-70): 1) press (shown here is the lightweight and portable Lee Hand Press), 2) Hornady New Dimension three-die set for .45-70, 3) shell holder (shown here is the RCBS #14), 4) Hornady/Pacific M powder scale, measures 0 to 509.9 grains, 5) gunpowder (two different rifle powders shown), 6) primers, 7) empty cartridge cases ready for loading, 8) bullets—one 50-count box of jacketed soft point, one 50-count box of lead flat point, 9) hand-loading manuals, 10) primer pocket reamer and Forster case de-burring tool, 11) powder funnel and spoon for scooping and pouring powder, 12) tumbler for polishing cases (Hornady M3 Case Tumber), tumbler media, and One Shot case lube, 13) loading block/tray for keeping loads organized, 14) bullet puller, inertia hammer type from Quinetics Corporation, in case you wish to safely unload any hand loads.

savings can be realized over the long run by those hand loaders who shoot often by reusing the same cases multiple times and focusing their investments on the other main loading components. Even so, the bullets, powders, and primers are not cheap, and the necessary loading tools are at least initially costly. Additionally, hand loading is a time-consuming activity, no matter how efficiently it is accomplished.

As already suggested, some shooters have minimized their cost per round by buying military surplus ammunition in bulk, often at wholesale prices. The quality of military surplus ammo depends a lot on its country of origin as well as its year of production. A lot of old ammunition can be corrosive to your gun barrel if not quickly and thoroughly cleaned following shooting, and its storage history will usually be unknown. Likewise, the quality control standards do vary from the factories of one country to another, and components such as steel cases and Berdan primers that are common in some foreign ammunition will not be suitable for reloading, as a general rule.

Additionally, buying anything in bulk requires a considerable investment, and military surplus ammo is limited only to the military-type cartridges. Nevertheless, the cost per round is indeed usually lower with surplus ammo than with commercial factory-loaded ammunition, so this is something for us to weigh.

Uniformity in firearms and ammunition can certainly be a beneficial goal for anyone (or for a group) shopping for an assortment of weapons to serve postapocalypse survival needs. The interchangeability of parts between firearms can simplify the weapon support requirements, and having a narrower variety of calibers to stock up on can save time, space, confusion, and money, as well as maximize consistency in sighting and weapons proficiency training.

It's difficult to generalize about the best place to store ammunition, because each shooter will have his own preferences and priorities, which may include ideas about accessibility, safety, secrecy, or even long-term preservation. Apartment dwellers will face different issues than homeowners, as will people living in cold versus humid climates, or gunowners with children to consider. It is generally recommended to store ammo in a cool, dry place protected from the elements. For some people, that will mean in a basement up on pallets. Others may prefer keeping all their ammunition in a gun safe. You have to assess your own situation and make decisions accordingly.

CHAPTER 7

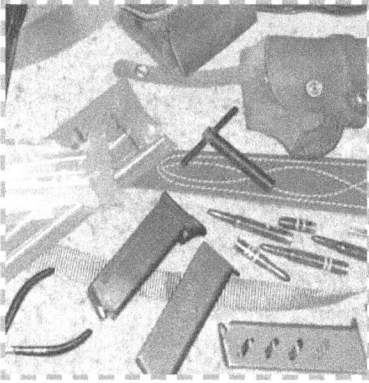

Weapon Support Gear for the Apocalypse

Choosing the best weapons that are chambered for the most ideal cartridges for the coming apocalypse is a monumental task in and of itself, but the decision making surrounding this task doesn't end there. We have other related things to consider, such as all the paraphernalia needed to use and maintain our weapons—things like extra magazines, cartridge holders, bandoleers, cleaning accoutrements and assembly/disassembly tools, holsters, protective shooting gear, gun cases, slings, bipods and tripods, spare parts, or any other gear that may be useful in conjunction with our doomsday weapons.

In my experience, a kit bag organized to support each firearm makes the weapons more convenient for use. Sometimes called a range bag, a firearm support kit bag will typically contain extra ammunition, cleaning and disassembly tools, optics, ear and eye protection, shooting rests, targets or target markers, and often spare parts for whatever weapon the kit supports. A decent container might be a durable gym bag with a camouflaged outside, or a small backpack or rucksack could serve the purpose.

Military snipers, I learned from an interesting video on the subject, call their support kit a "drag bag." Within it the sniper stows things like extra ammunition, a spotting scope or binoculars, laser range finder, bore and chamber cleaning tools, laminated bullet trajectory charts, and whatever else might be needed to support the weapon or help the sniper execute his mission in the field.

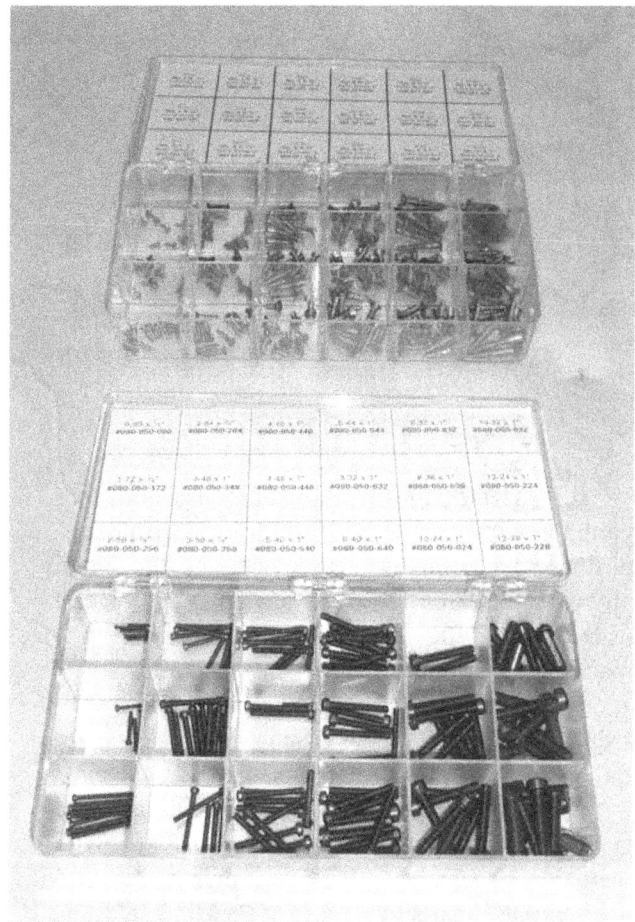

Handy gun screw kits from Brownells. One includes machine screws and the other wood screws, mostly for screwing items to gunstocks.

Gun owners invariably accumulate an assortment of shooting supplies to support their firearms.

SPARE PARTS

It is essential to stockpile an assortment of spare parts for any weapons selected for long-term or postapocalypse survival. As devices age, they gradually wear out, especially after any degree of regular usage. We are all witnesses to this process. We see our machines, tools, musical instruments, home appliances, and especially our automobiles become worn, require preventative maintenance, and occasionally fail. It is the same with firearms.

It is easy to lose pins, screws, springs, and other small parts during cleaning or disassembly for field maintenance and, being thinner or lighter, they often are the easiest parts to break. For these two reasons, such spare parts are the first I acquire in quantity for my weapon support bags.

Along these lines, I suggest you obtain an

The firing pins of CZ-52 pistols are notorious for breaking. It is a good idea to keep extras for this pistol.

The canvas bag at lower right contains rifle cleaning tools, bore light, ear muffs and shooting sunglasses, optics (binoculars, in this case), extra magazines, one stripper clip, ammunition, literature pertinent to the weapon type, disassembly tools like hand-guard clip pliers and gas cylinder wrench, plus a comprehensive assortment of spare parts in a sealed plastic bag, all to support this M14 rifle at the top. This is a great way to keep any weapon ready for the apocalypse.

exploded-view illustration and maybe step-by-step assembly/disassembly instructions (or perhaps the owner's manual) for your particular model of long-term survival firearm. Either seal it in a clear, water-proof plastic bag, or laminate the individual pages and store them with your spare parts. I laminate other paper reference charts and data as well, which makes these important items more durable and less vulnerable to the effects of wind, rain, or snow.

You might not see yourself as a gunsmith now, but you may be forced into that role in an apocalyptic future, when trained gunsmiths will be hard to find. In that case, now is the time to add a couple reference books on basic gunsmithing to your supplies to supplement the information in your gun owner's manuals. Amazon lists a dozen or more titles on gunsmithing, but one that addresses the unique requirements of repairing firearms under primitive conditions is *Guerrilla Gunsmithing:*

Quick and Dirty Methods for Fixing Firearms in Desperate Times by Ragnar Benson (available from Paladin Press).

As it is with all but the most complex of do-it-yourself projects, typical firearm maintenance and repair jobs are a matter of following a step-by-step process. Follow the instructions, work slowly and methodically, and you should have no trouble handling these essential chores on your own.

CLEANING AND MAINTENANCE GEAR

Give special consideration to the cleaning gear for your chosen weapons. A complete cleaning kit appropriate for the calibers you are using would be ideal, space permitting. The vast majority of regular shooters will already have most of the gun cleaning supplies they normally use, but the apocalypse kit warrants the most comprehensive, well-stocked kit

The comprehensive gun cleaning kit housed and well organized in a large fishing tackle box.

you can assemble for the long haul. It could be called upon to serve more diverse chores than you might otherwise encounter on a regular basis—everything from extracting stuck shell casings from chambers or scrubbing out dirty rifle bores, to changing sights and eventually repairing broken parts. Additionally, resupply of bore brushes, solvents, oils, and other items could become more restrictive after the gun shops are closed, so plan accordingly.

If you have the space for it and also the time (and money) to put it together, a comprehensive cleaning/maintenance kit will become enormously useful to keep all your weapons operational and ready. I found that a large fishing tackle box conveniently houses and protects the kit I assembled. It includes numerous screwdrivers, pliers, bore lights, bore brushes, bore patches, cloth rags, assorted jags, cleaning rod sections, small maintenance tools, picks, solvents and oils, pull-through cords, and other handy odds and ends.

The trays have little compartments that keep the small items well separated and organized. I didn't put together this whole kit all at once but instead simply added items to the box over a period of time. That way I never really noticed the costs of its contents.

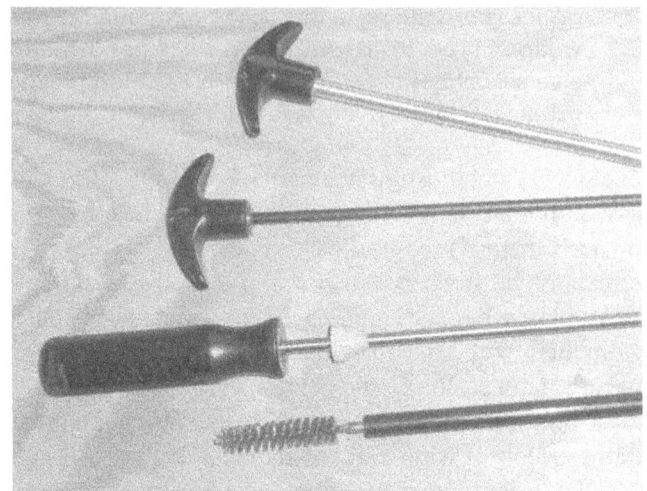

Cleaning rods of different materials: aluminum (top), steel (the two in the center), and fiberglass (bottom).

For something you might be more likely to toss into your emergency bug-out bag intended for foot travel, consider keeping a smaller and more portable range kit containing just the main essentials for maintaining one or two specific firearms. You can conveniently house

This small gun cleaning kit includes: 1) bag to house everything, 2) steel takedown cleaning rod, 3) ribbed and slotted cleaning jags, 4) rifle/pistol bore brushes, 5) shotgun bore brush with thread-size adapter, 6) toothbrush-style gun brush, 7) pull-through bore cleaner rope, 8) bore guide to center steel cleaning rod, 9) pull-through bore brush, 10) package of pipe cleaner brushes, 11) WD-40 cleaning solvent and lubricant, 12) fiber-optic bore light, and 13) cloth rags and patches.

such a kit in a shaving kit travel bag or similar container. This will be the kind of gun cleaning kit you would most likely have on a hunting or camping trip.

You will want an assortment of cleaning rods of several sizes for pushing brushes and patches through gun bores, as well as for pushing out objects that occasionally get stuck inside bores or chambers. Brass, aluminum, and hardwood cleaning rods/ramrods will cause less damage to the insides of gun barrels but naturally won't hold up as well as steel. Fiberglass, on the other hand, is very durable, won't rust, and won't harm the muzzle's crown or the rifling inside the barrel. For an apocalypse scenario, I recommend fiberglass.

Pull-through cords with cloth patches for routine cleaning or to remove snow, mud, dirt, cobwebs, powder residue, excess grease and oils, or other gunk from barrels are lightweight and easy and cheap to make.

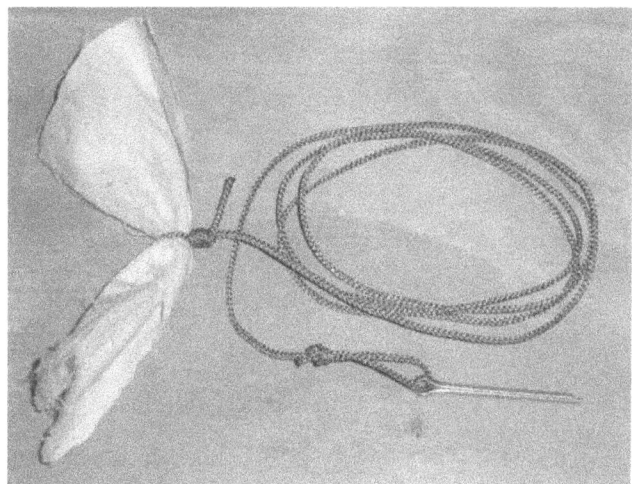

Pull-through bore patch on a string, with cotter pin for weight on the other end that drops down the barrel. Easy and quick to make and use.

Firmly cinch a strong, small-diameter braided cord with a hangman's knot in one end around the middle of a rectangular cloth patch. The hangman's knot makes it convenient to loosen after use to switch from a dirty to a new, clean patch. A small weight tied to the other end helps it drop easily down the barrel; a heavy cotter pin seems to work well, as will lead fishing sinkers. I keep several of these cords in my gun cleaning kits.

Pull-through ropes for cleaning gun barrels have been sold commercially for at least 10 years. Hoppe's calls its product a BoreSnake, and they advertise it as the "World's Fastest Gun Bore Cleaner." The cord has a brass weight at one end, and brush bristles are embedded in the rope to loosen hard deposits. The main floss area in the rope is described as having 160 times more cloth area than a patch. These products are compact, lightweight, and available for all common calibers. Most importantly, they are washable and reusable, which makes them ideal for long-term use after patch supplies are depleted.

As far as firearms maintenance tools go, you will want at least one good set of gunsmith screwdrivers. They are hollow ground as opposed to the general-purpose tapered slotted type, and they are more propriate for use on the heads of the precision machine screws found in guns. This is important, because you don't want to strip the screws on your firearms when access to replacement parts will be limited or nonexistent.

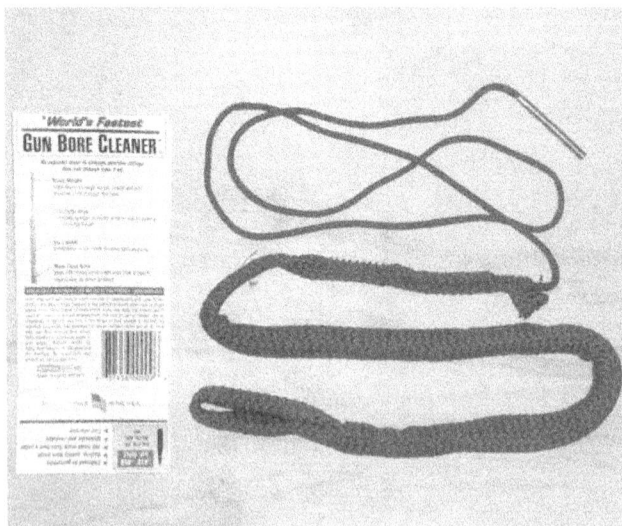

Chapman Tool's compact gunsmith screwdriver set is perfect for the gun maintenance kit.

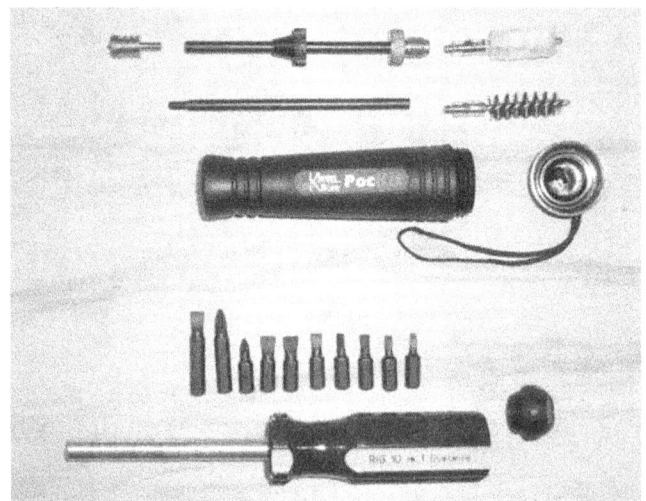

RIG 10 in 1 Gunsmith screwdriver kit at bottom, with 10 screwdriver tips stowed in the handle. Kleen Bore's PocKit at top is a similarly handy tool for bore cleaning, with components also stowing inside its handle.

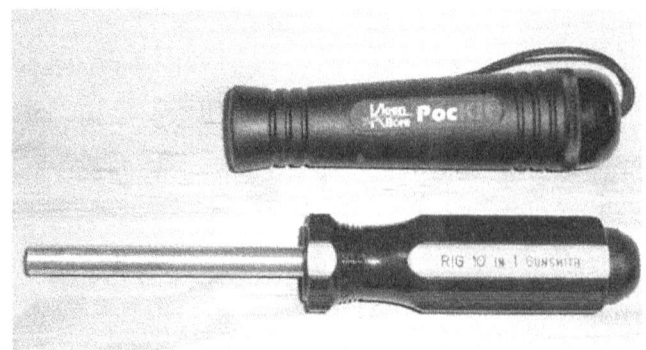

A bore snake for cleaning .44 and .45 barrels. This particular product was produced by National Tech-Labs in Boise, Idaho.

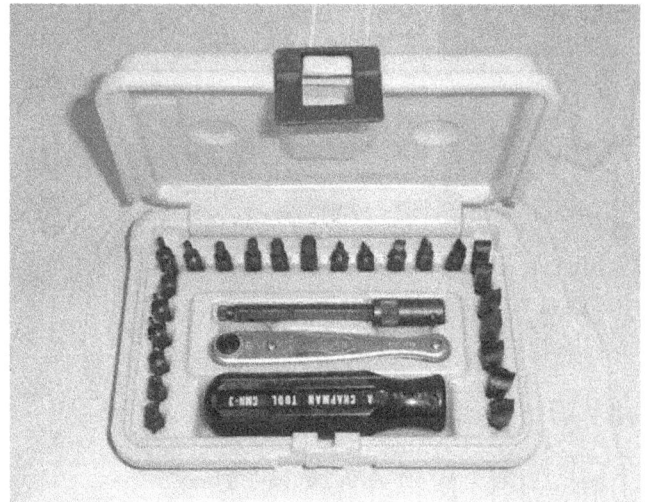

Both kits conveniently contained in their handles.

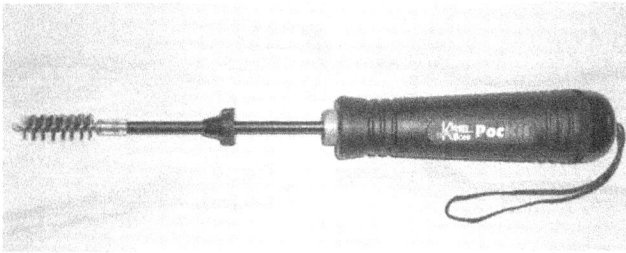

The PocKit is ready to clean a handgun barrel.

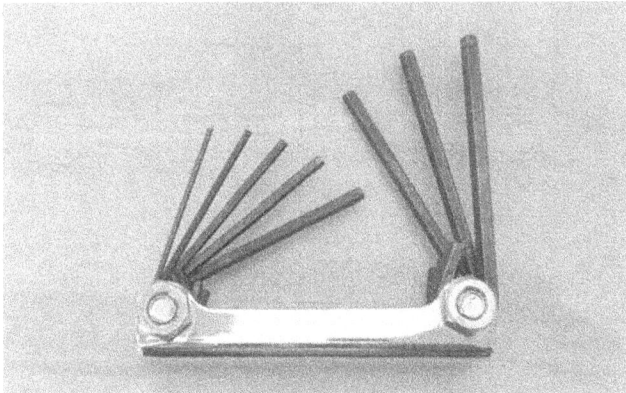

This handy eight-tip Allen wrench set from Eklind Tool is worth keeping in the firearm support kit.

A set of small brass and steel punches, wooden mallet, hardwood dowel for moving the soft plastic rear sights on GLOCK pistols, and small brass hammer all find a place in the kit.

A set of the smaller sizes of Allen wrenches will be useful for changing and adjusting scope mounts, certain sight configurations, sling swivel studs that clasp to barrels or magazine tubes, accessory rails, and the numerous other attachments that require them.

An apocalypse gun maintenance kit will be more complete with at least one pair of precision needle-nosed pliers and perhaps a set of tweezers for gripping tiny screws and pins in narrow spaces. A broken shell

An assortment of lubricants, gun cleaners, and bore solvents should be part of every shooter's kit.

extractor for every caliber could also be enormously problem-solving when needed. They cost around $10 apiece and are available from Brownells.

Drift punches, especially brass punches of various sizes, will come in handy whenever you need to knock a pin out, push a small part within a narrow space, or drift a sight base that is dovetailed into the barrel to one side or the other. Brass is the ideal material for gunsmith punches because it is softer than steel and won't ding or dent the surface of steel gun parts the way steel punches will. A small hardwood dowel can be used as a punch on certain delicate or soft plastic parts, such as the rear sights on GLOCK pistols. A wooden mallet or small brass hammer will also be useful in conjunction with the punches.

I often use WD-40 as a kind of general-purpose gun cleaner, even though it has a tendency to collect dust more than a thinner machine oil is likely to do, and it seems to gum up after a while in my experience. But unlike other gun oils, machine oils, solvents, and the majority of specialty gun cleaners on the market that I am aware of, WD-40 will function as both a solvent *and* a lubricant, so you can use it for both purposes at the same time. In spite of its deficiencies, WD-40 is a versatile product that can also be used for other household purposes like removing sticky labels from glass jars.

Cleaning lead deposits out of a barrel's rifling grooves can be quite a chore sometimes, for those of us who occasionally shoot cast bullets. It has been my experience that the common brass wire bore brushes, even those that correspond in size to a given bore diameter, rarely brush aggressively enough to scrape out all the lead. Consequently, I started a habit of using oversized brushes, like a .38-caliber brush in a .32-caliber barrel, or a .45 brush in a .40/10mm barrel, and so on.

This is hard on the brushes and it doesn't usually take very long for them to become deformed beyond saving, but they tend to scrub out the barrel faster and more thoroughly. Shooters can never really have too many brushes in their cleaning kits anyway, in my opinion.

RIFLE RESTS

The most accurate shooting requires the steadiest platform to support and stabilize the weapon for every shot. Shooters have used every kind of support or rest imaginable to aid with this. I can think of at least two practical possibilities that are both lightweight and compact: 1) cross-sticks, and 2) the forearm-mounted bipod.

Cross-sticks were used extensively by buffalo hunters on the Great Plains in the nineteenth century, when large-bore rifles were used for taking shots at long ranges. Wooden sticks commonly served this purpose in the past, and a pair of sturdy dowels of the

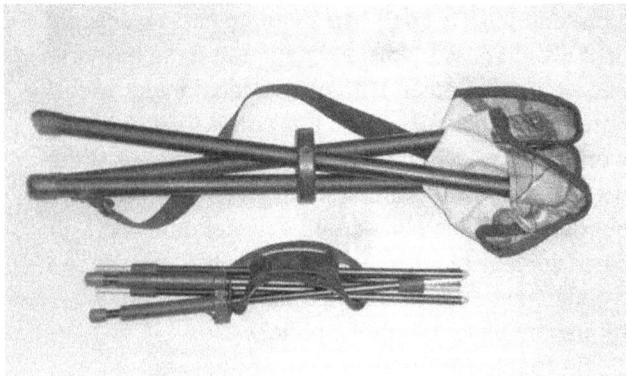

An inexpensive, collapsible, cross-sticks type of bipod that weighs only a few ounces can be a useful item to carry in a hunting pack, along with a portable tripod seat.

Using the lightweight cross-sticks and tripod seat.

The inexpensive bipod attached to the sling swivel stud of this rifle makes it very steady for shooting in a prone position.

desired length lashed together with a bootlace just below the functional V-notch rest could serve as improvised cross-sticks to steady any rifle.

Nowadays in sporting goods/hunting stores, you can find inexpensive, lightweight, plastic cross-sticks that collapse just like tent poles into a compact unit for stowing in a hunting pack. Even though it appeared rather flimsy at first, I found that this arrangement works fairly well in conjunction with a small, foldable tripod seat that allows the shooter to sit at just the right height while resting the forearm of his rifle. The little tripod seats are themselves worthy products for the apocalypse survivor, also being conveniently portable. If nothing else, the seat will help you avoid sitting on the cold, wet ground outdoors. If you are forced to hunt for survival, you'll have enough challenges to face without adding hypothermia to the list!

Military snipers and sportsmen alike have used forearm-mounted adjustable bipods on their rifles for years, and this is a logical accessory for any bull-barreled, heavy scoped rifle. There are a number of bipods on the market, and prices range from around $35 for the cheapest imported models to several hundred dollars for some of the high-end tactical units.

I discovered that even the least expensive bipod makes a noticeable difference in how quickly and solidly a rifleman is able to steady his weapon while shooting from a prone position. I attached the XLA 6- to 9-inch model from Caldwell Shooting Supplies (made in China and under $40 retail) on my Remington Sendero, and it is completely functional. The feet are easy and quick to extend or retract, and the unit is adequately stable. What I like best about this system is

Adjustable bipod mounted to the forearm sling swivel stud on a Remington Sendero, folded up for transport and storage.

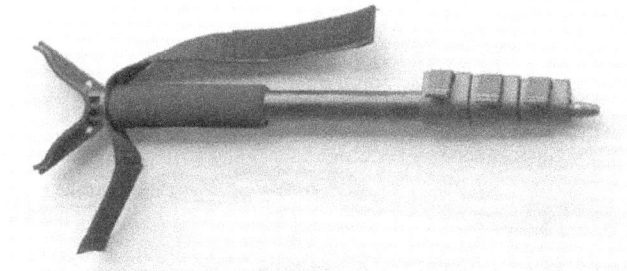

The Compact Shooter's Rest from the Allen Company, collapsed for stowing in a pack.

The author's son uses the single-leg rifle rest to support a rifle from a seated position.

the way it mounts to the rifle's forearm: it clamps to the forearm sling swivel stud, and the sling and swivel can then be attached to the bipod's own swivel stud. Attaching or removing the bipod from the gun takes less than two minutes without tools, and there are no modifications or screw holes necessary to the weapon. Whenever I wish to upgrade to a more expensive model, it should only take me a minute or two to switch them on the gun.

Harris Bipods are very popular, and the company offers a variety of bipod-related products (www.harris-bipods.com). Another source for high-end military bipods is GG&G Tactical Rifle Accessories (www.gggaz.com).

If you want something even more compact for field use, the single-leg rifle rest might find a place in the shooting kit. This collapsible Y-shaped pole can provide quick, convenient support to stabilize a hunting arm in the woods, where a more elaborate rest may not be available.

I bought one of these telescoping shooting rests recently at the local sporting goods store for $15. Weighing just 6 ounces, the unit is composed of plastic and

thin aluminum and extends from a compact 15 inches to 3 feet in length. Its four telescoping segments can be set at various heights that, even when fully extended, can be locked into a fairly rigid support pole. Other models on the market will provide even taller support.

Do not underestimate the utility of a rifle rest in your cache of accessories. Ammunition will become precious after an apocalypse, so *every* shot needs to count. You need to do everything you can to maximize your chances of a hit, and these rests are ideal for accomplishing that.

SLINGS FOR LONG GUNS

One of the first things I do after acquiring a new long gun is attach—if the weapon doesn't already wear it—a properly fitting set of sling swivels and a sling. This is one accessory that I consider almost fundamental for just about every carbine, rifle, or survival shotgun, because it can make carrying and supporting the weapon so much easier in the field. Remember, there could come a time when gasoline is unavailable, so throwing the gear in the vehicle and heading out to the woods will no longer be an option. You may have to tote everything you'll need.

There are variations of the basic rifle sling that are used for different specialty applications. Competitive rifle shooters, for example, might use a special single-point sling as an aid to steady the rifle rather than as a carrying aid, while military and law-enforcement personnel might have need for a three-point configuration that functions like a harness, allowing the shooter to release his grip on the weapon without dropping it. Or the innovative Ching Sling system might be employed in conjunction with the Scout Rifle or other tactical carbine as both a carrying strap *and* an aiming support aid. The point is, a sling of one version or another will be infinitely utilitarian, and anyone preparing for the coming apocalypse will probably want some variation of a carry sling on his weapon.

Perhaps the three most important features for our purposes are that the sling be comfortable to use, adjustable, and durable. A sturdy leather sling could serve the purpose and it can be an attractive accessory, but the right synthetic product will tend to be more impervious to moisture and require less maintenance. The one advantage of a leather rifle sling for the apocalypse is that a functional version is fairly easy for someone with any leather craft experience to fabricate from just a strip of 9- or 10-ounce cowhide, a leather punch, and some leather lacing. The product can then be wiped down with neatsfoot oil from time to time to help preserve it.

The wider the sling, the better it distributes the load over the surface of the shoulder and the more comfortable it can be expected to be. This could be an important factor in a situation involving long-range foot travel.

If you are ever forced to go with a shoulder weapon that wears no sling and you have no sling available, you could always improvise a usable substitute. A length of ½- or 5/8-inch diameter rope can be used as a functional temporary sling in a pinch. Such a rope sling for a rifle or shotgun is quick and easy to makeshift. It works well with a constricting hangman's knot at both ends of the rope that cinch up firmly around the weapon wherever desired.

HANDGUN HOLSTERS

Holsters are not essential gear for the apocalypse. Handguns could be securely tucked under a waist belt, toted in a coat pocket, or stored in a fishing tackle box.

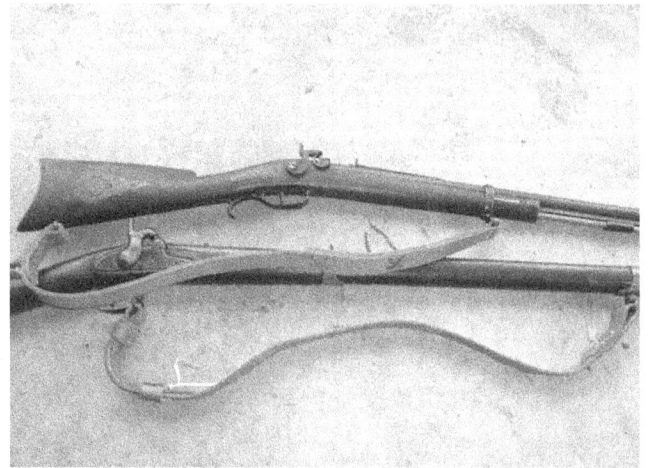

Two crude but functional leather rifle slings fabricated by the author.

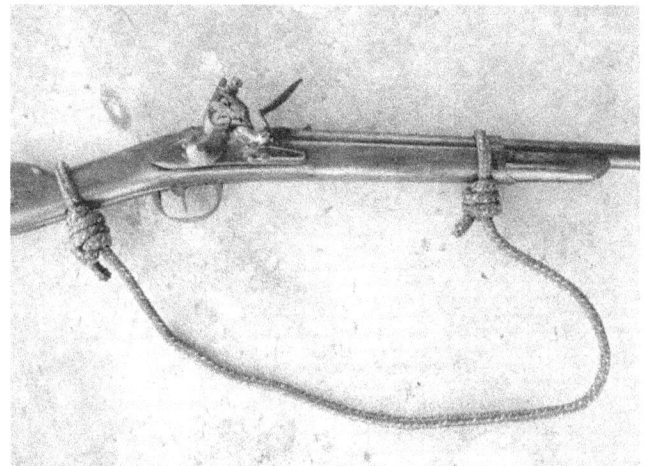

A length of half-inch diameter rope is secured to this flintlock carbine with a hangman's knot at both ends. The end attached to the forearm is prevented from sliding up by the barrel band.

However, most of the survivors of the apocalypse who opt to pack handguns will surely want holsters for the task, and the seemingly endless variations on the market can be confusing for someone trying to decide on the most suitable carry rig.

The first place to start the holster selection process is with the decision about the specific models of handguns to be carried so that the holster is chosen specifically for the correct handgun it was made to fit. The second step is to try out the different holster designs at the local gun store to determine which one is most comfortable for you. Rigs are available for carrying handguns on the hip in a waist belt, under an arm in a shoulder harness, inside the pants, in belly bands, inside pockets, or on the ankle. Each shooter will

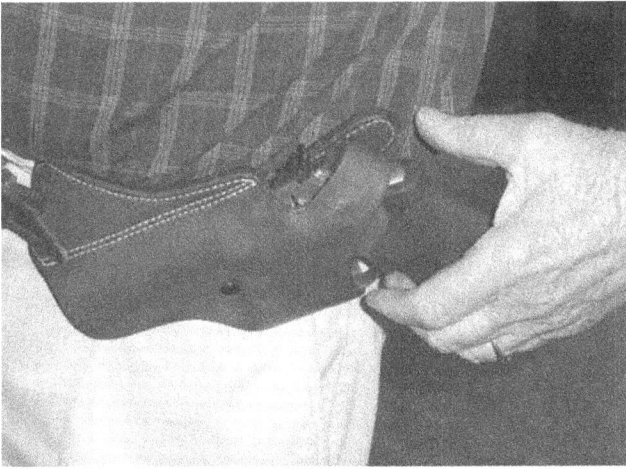

This versatile leather revolver holster can slide onto a waist belt in any of several positions, depending on the shooter's gun-wearing preference.

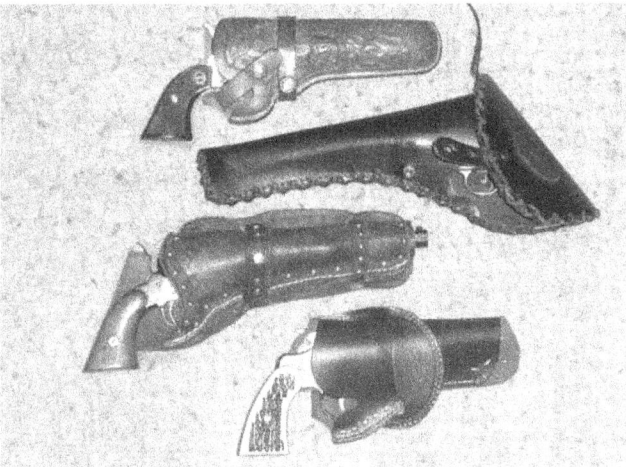

Several common leather revolver hip holster designs. Nothing fancy, but perfectly functional.

Kydex holsters are durable and perfectly fit for many popular handguns (here a Springfield XDM 9mm).

Nylon holsters might be more practical for the apocalypse.

develop his or her personal sense about what is comfortable, practical, or affordable. Prices for new holsters range from under $20 for generic hip holsters to well over $100, depending upon the brand, level of quality, design, style, popularity, and materials of composition.

I favor different holster designs for different applications, but when selecting products for a postapocalypse era, I would lean toward products comprised of synthetic materials over leather, simply because the synthetic materials are generally less affected by moisture and require less maintenance than do the leather products. There are some excellent leather holsters on the market that would surely serve survivors quite well, but I recommend giving synthetic models serious consideration. The rigid and very durable Kydex (thermoplastic) handgun holsters, for example, have become incredibly commonplace in recent years, especially with law enforcement. They are simple, rugged, and perfectly molded to the various popular gun models.

While you're at the gun store shopping for suitable handgun holsters, it might also be a good time to obtain padded or hard-shell gun cases to protect those doomsday long guns. Gun cases aren't essential items, but they can make the long guns much easier to hand carry or transport in a pickup bed or trunk of your car, and they do provide a degree of protection for the weapons.

CARTRIDGE BANDOLEERS

Something as seemingly insignificant as a cloth or leather cartridge bandoleer could actually become a critically important piece of gear after the apocalypse. Not only does this item provide a useful means to keep loaded rounds conveniently accessible like no other system, but a cheaply constructed version that fails to retain the cartridges securely could have you spilling ammunition as you run for your life through the woods, wasting your precious ammo as well as leaving behind a trail of shiny brass by which to track you. This is one piece of firearm support gear that merits paying for quality.

Just as with rifle slings and holsters, leather bandoleers look nice when new, and they are certainly usable for short-term applications. But leather requires special maintenance to prevent drying and cracking, does not age or weather particularly well, will eventually stretch, and reacts with brass over time to form a greenish, gummy buildup that can be difficult to remove. Likewise, some of the thinly constructed synthetic cartridge loops tend to lose much of their elasticity after a while and often degrade like old rubber bands, their minimally sewn connections eventually separating or unraveling at the worst possible moments.

When shopping for bandoleers you expect to be usable after the apocalypse, look closely at the quality of the product. It's hard to beat thick ballistic cloth having heavy stitching. And I will always opt for an American-made product over anything imported from Asia.

About 10 years ago I bought a bandoleer with a

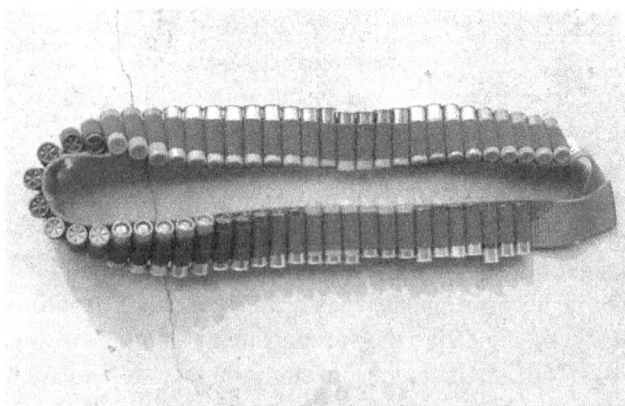

High-quality 56-round bandoleer for shotgun shells from Brigade Quartermasters.

Brigade Quartermasters label (www.brigadeqm.com) that holds 56 12-gauge shells, features heavy-duty construction, and seems very durable overall. I have stored it with shells in the loops since I bought it, and it has not yet lost any of its firm gripping capability. The product still looks and functions like new.

OPTICS

Hunting, target shooting, patrolling, scouting, surveillance operations, and especially shooting across long distances are all tasks that can be accomplished much easier—and often more successfully—with the help of optical vision aids, or optics.

Binoculars are excellent firearm support optics because they are quick and easy to employ, relatively compact, and provide useful magnification for viewing targets. When shopping for the right pair of binoculars for the apocalypse, give special consideration to the product's optical clarity and brightness, magnification power, size and weight, and overall durability.

Optical clarity and brightness will be influenced, as is the case with any type of optic having magnification, mainly by the size of the objective lens. This is because a larger objective lens lets in more light than does a smaller one.

When you see numbers like 8x40, that first number indicates the power of magnification. In this case, the image viewed through the optic will appear eight times larger than the same image viewed from the same distance with the naked eye. That second number, 40 in this example, indicates a 40mm objective lens. Thus, a 10x40 scope or binoculars will have greater magnification than will an 8x50, but the objective lens of the 8x50 lets in more light, so the image will (in all likelihood) appear brighter and possibly even clearer than the 10x40, even though its magnification is not as powerful. Such other variables as lens designs and innovative features will also play a role in the brightness and clarity of the optic, but the amount of light it captures will probably be the largest contributing factor.

For our purposes, we will want the best of both worlds. You will surely want magnification for identifying hard-to-see targets as far away as possible, while at the same time viewing them with the utmost clarity. Ultimately you will have to decide for yourself which one of these two benefits, clarity or magnification, you are willing to give up some in order to gain more of the other. When both numbers start getting

From left to right: Simmons 8x40 rubber armored binoculars, Tasco Futura SE 10x42 fully coated, and the compact Brunton Lite-Tech 5038W 10x25 binoculars. Any of these products would give survivors a visual edge over those without optics.

larger—especially the millimeter size of the objective lens—the physical size of the binoculars also gets larger, because the lens requirements are more substantial. Just take a look at the 10x50 binoculars at your local sporting goods store; every brand I am aware of is considerably larger and heavier than models with smaller objective lenses.

Now that we've talked about clarity, magnification, and the size of the unit, let's talk about durability, because survivors of the apocalypse are going to need rugged equipment that can endure the harshest environments.

The high-end binoculars on the market today are well-built instruments that are normally very rugged. Many are advertised as being waterproof and fog proof. Coated lenses that reduce reflection and glare are popular. I also like rubber-armored binoculars, because this gives them a higher degree of shock resistance and provides a more solid, comfortable grip.

New binoculars range from under $100 to almost $3,000 for the best models made by such companies as Leica and Swarovski. But such popular brands as

Leupold, Nikon, Redfield, and Bushnell offer good binoculars for less than $300. In the end, you get what you pay for, but every pair of binoculars I own cost under a hundred bucks, and they have all served me well.

A spotting scope is an incredibly useful piece of equipment whenever doing any kind of target shooting or long-range observing. If the apocalypse kit has room for one, it could be well worth its weight in gold at some point.

The spotting scope is a telescope well suited for use on shooting ranges, when you want to view where the bullets are hitting on the target without having to move closer to see. The magnification power is often much greater than what you will find with binoculars, and variable magnification and manually adjustable focus features are common. With their higher magnification, spotting scopes are typically used in conjunction with a tripod to help steady the image, although they can also be hand-held by someone with well-supported elbows.

Range finders, especially the laser-type units com-

This Bausch & Lomb spotting scope can be hand-held, as shown here, or used with a tripod.

Bushnell's Yardage Pro 1000 laser range finder.

monly used by hunters and shooters today, make it remarkably convenient to determine distances accurately. Laser range finders operate by sending out a laser beam pulse to the target and measuring the time interval of the pulse's return after being reflected by the surface of the target.

Even though I am not a huge fan of battery-dependent devices for any long-term survival requirements, I almost want to make an exception for the laser range finder, considering how important such a piece of equipment could ultimately be. With an operational

unit, you could stand in one position and determine the distance in either yards or meters to any stationary object that will reflect the beam back up to several hundred yards/meters away. Current models typically read out to 400 yards or more with just a click of a button. My range finder uses regular 9-volt batteries, which you could easily add in quantity to your pre-apocalypse stockpile of supplies.

EAR AND EYE PROTECTION

A complete shooting kit will always contain equipment for protecting the shooter's ears and eyes. Continuously exposing the eardrums directly to the sharp reports of centerfire weapons will eventually take its toll on a shooter's hearing. Even firing one round from a .50 BMG rifle without wearing any ear protection can permanently damage the eardrums. Hearing protection is provided to a degree by inexpensive earplugs, but shooters earmuffs are even better. Different levels of protection are available, depending on the product.

The eyes are also vulnerable whenever firearms are discharged close by. Empty cartridge cases get ejected from automatics and fly in unpredictable directions, hot gasses occasionally escape from loose actions or from between revolver cylinders and barrels, and rifle scopes sometimes rocket back into the shooter's face during recoil. Wearing durable shooting glasses or sunglasses could potentially prevent damage to a shooter's vision.

Ear and eye protection will be important to anyone who plans to target practice.

SNAP CAPS

Snap caps should probably be included with the firearm support gear because with them, dry firing the weapons is less problematic. They are designed to cushion the strike of the firing pin with their spring-supported (or rubber) dummy primers, thereby saving firing pins from excessive shock. Dry firing weapons can be a practical way to practice proper shooting techniques without wasting live ammunition—an important consideration for a time when every bit of ammo will be precious. Also, with such firearms as striker-fired GLOCK pistols, dry firing the weapon is a step in proper disassembly. Snap caps are available for nearly every common caliber, including shotguns and even .22 LR rimfire weapons.

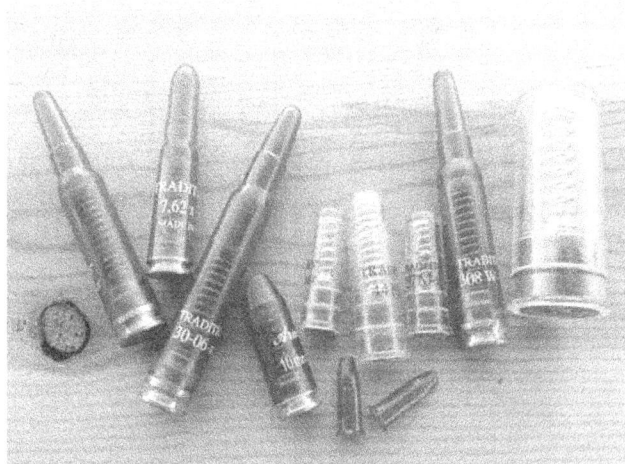

Snap caps for various centerfire and rimfire weapons.

CHAPTER 8

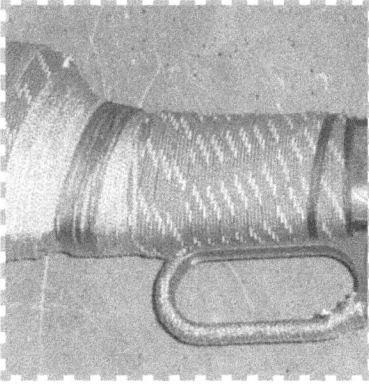

Practical Modifications for Apocalypse Weapons

Ideally, whatever weapons we select for our long-term survival should be usable just as they come from the factory. But here we will consider a few modifications that might make some of them even more suitable for various desperate survival scenarios.

It is imperative that any modifications you make to your chosen apocalypse weapons do not impair their function or render them dangerous in any way. It is also important to avoid any modifications that would turn your weapons into illegal weapons. It's always nice to stay out of jail whenever possible.

There is also the potential of lowering the resale value of a gun as a result of personalizing it. Each individual must decide for himself the extent to which he would be comfortable adversely affecting his monetary investments. Not being a gun dealer or even a collector of valuable firearms, I am more concerned with the practical utility of my guns for survival purposes. I see them as one of the many tools that I might need in an emergency.

One clear advantage we have in our time is the sheer volume of aftermarket components for some of the more popular modern firearms. The different ways to alter or customize the basic 1911 pistol, for example, are simply beyond listing here. The same is true with such other popular firearms as Ruger's 10/22 semiautomatic .22 rifles, the tactical AR-15-type rifles and carbines, and to a large degree the GLOCK pistols (see, for example, *How To Customize Your Glock* by Robert Boatman and Morgan Boatman, published by Paladin Press).

Two of the largest sources of gun parts in America are Numrich Gun Parts Corporation (www.gunpartscorp.com) and Brownells (www.brownells.com). Brownells is an excellent (though expensive) source of new parts for modern firearms, as well as gunsmithing tools and all kinds of shooting paraphernalia, while Numrich always maintains a vast inventory of parts for old and obsolete firearms.

OPTIONAL SIGHTS FOR APOCALYPSE WEAPONS

Multiple sighting options can often be incorporated with a particular weapon, adding to its versatility. One example of this is the rifle with a scope mounted on open rings that will allow the shooter to alternatively view the gun's open sights under the scope. Another example is the handgun with a laser sighting system that allows use of the weapon's conventional sights as well as the laser. Sighting arrangements like these, if we are to view them as a form of modification, give shooters multiple options and increase the versatility of their tool.

It might be useful at this point to briefly review some of your sighting options before considering any modifications to your weapons. The simplest firearm sights—as well as the most durable, least expensive, least obtrusive, and lowest maintenance—are the various styles of open sights. We're talking about a simple V-notch or squared U-shaped groove over or near the receiver used in conjunction with a blade or post front sight above the muzzle. The front and rear sights are aligned together with the target by the shooter's eyes during aiming.

Open sights can be adjustable for windage (side to side movement) and elevation, or fixed (nonadjustable). Variables that affect the success of open sights include the sight radius (distance between front and rear sights), their shape and dimensions (finer blades and notches tend to provide more refined sighting), and their visibility for the shooter's eyes.

Aperture, or "peep," sights represent one viable alternative to the conventional V-notch open sight. With this system, the shooter's eye aligns the front sight through a tiny peephole in the ringed rear eyepiece during aiming. Centering the front sight post inside an aperture ring is natural for the eye, which seeks an equal amount of light around it. Peep sights are popular with target and match shooters in certain competition categories because of the ease with which they can help achieve superb accuracy.

A less precise, though quicker to use variation of the traditional peep sight is the popular ghost ring rear sight, which simply provides an eyepiece having a considerably larger aperture within which to center the front sight during aiming. Ghost rings are now quite common on multipurpose shotguns and handy carbines used mostly for close-range hunting and tactical applications.

Shooters who desire greater visibility with their sights may wish to add night sights or any of the other innovative sighting systems to their weapons. Several interesting options include tritium night sight inserts, fiber optic sights, and red dot sights.

Tritium night sights are comprised of tiny glass capsule inserts containing radioactive gas that emits electrons, causing phosphors to glow and creating self-powered betalights. The tritium isotope has a half-life of less than 12½ years, meaning that the glow will progressively dim as time goes by. Years ago, I had tritium inserts installed on a revolver, and they seemed fairly bright for the first year or two. But it wasn't too long before these sights became noticeably less visible, in both daylight and low-light conditions, than the standard white outline rear and orange plastic insert front sights they replaced. Seven-year or older tritium sights will likely be of limited value, I believe, if they still have any usable glow to them at all.

Fiber optic sights are popular nowadays with bow hunters as well as shooters. These provide brighter than standard sight points by collecting available light from the surrounding environment and then directing it to the shooter's eye. Unlike tritium night sights, fiber optic sights can do what they do indefinitely. However,

An example of a red dot sight.

they can't gather any light from the dark, so they will not function as night sights.

Red dot sights have also become very popular in recent years. They use an LED (light-emitting diode) in conjunction with a special mirror to generate an illuminated red dot as an aim point that remains in alignment with the weapon regardless of eye position to facilitate quick targeting. Red dot sights are battery-dependent, but power consumption is very low and battery life can endure for thousands of hours.

EXPEDIENT SIGHTS FOR APOCALYPSE WEAPONS

I plan to obtain a barrel sleeve chambered for .30-30 Winchester to insert into a single-shot 12-gauge shotgun barrel, allowing it to be used as a rifle. To facilitate the project, I decided to add sights to the shotgun's barrel. Prior to starting the work, the shotgun featured only a bead on top of the muzzle end of the barrel for sighting.

This was my dad's idea, and I have to admit that it seems like a practical one, because the rifle sights should not impede the shotgun's function as a shotgun. The barrel of this particular shotgun is only a minimum legal 18 inches, so making a carbine out of it constitutes no great sacrifice as far as I am concerned.

Since shotgun barrels are typically thinner than rifle barrels, the logical method for attaching the sights

An open rear sight soldered onto the shotgun barrel near the action.

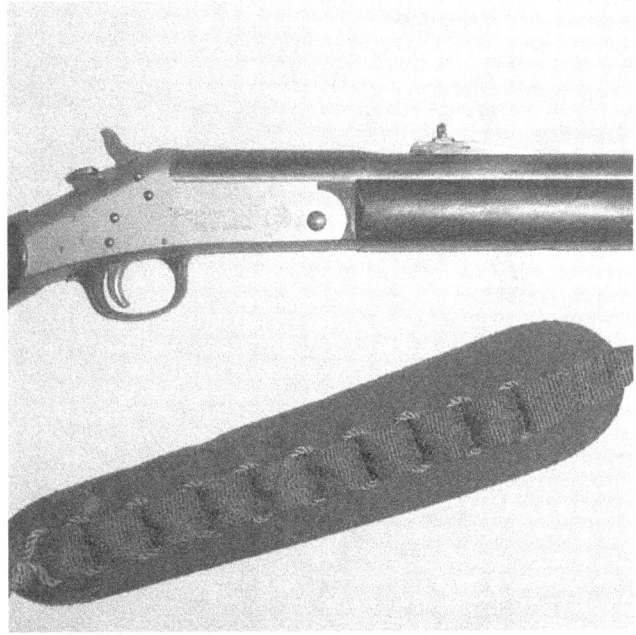

The single-shot shotgun with the rifle sights soldered onto the barrel.

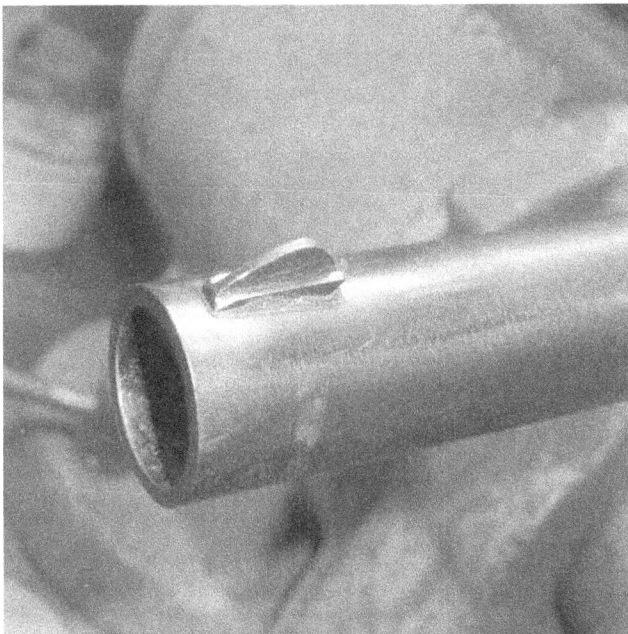

The front sight blade soldered onto the end of the barrel, after the shotgun bead was removed.

SHOULDER STOCK POSSIBILITIES

Rifle stocks and handgun grips can often be modified to make them more suitable for certain situations. Folding stocks are available for certain carbines and shotguns that make those weapons more compactable in a hurry, should it become desirable to stow the arm in a rucksack or some other confined space for convenience. Similarly, telescoping adjustable shoulder stocks are very popular now with AR rifles and carbines, allowing quick and easy adaptation to any situation.

With certain handguns, it might be possible to install some type of custom mount onto the grip handle that would facilitate attaching a shoulder stock, making the firearm functionally more like a carbine than a handgun. The reader is advised to research any applicable laws concerning this type of adaptation before proceeding to alter an expensive firearm and risk committing any possible violation.

Accessory rails on long guns and handguns have become really popular recently, because they allow such a high degree of versatility with a sort of modular adaptability. Flashlights, laser sights, telescopes, bipods, and all sorts of handles have been made for quick attachment on accessory rails. This can be a wonderfully useful feature for anyone arming for the apocalypse.

seemed to be silver soldering. Sights silver-soldered to a barrel will normally adhere tenaciously. We first shaped the sight bases to the contours of the gun's barrel, tinned them with solder, decided where they should be positioned, and, with the barrel secured in a vise, soldered both sights in place with a gas torch. Now I can hardly wait until I get that .30-30 insert!

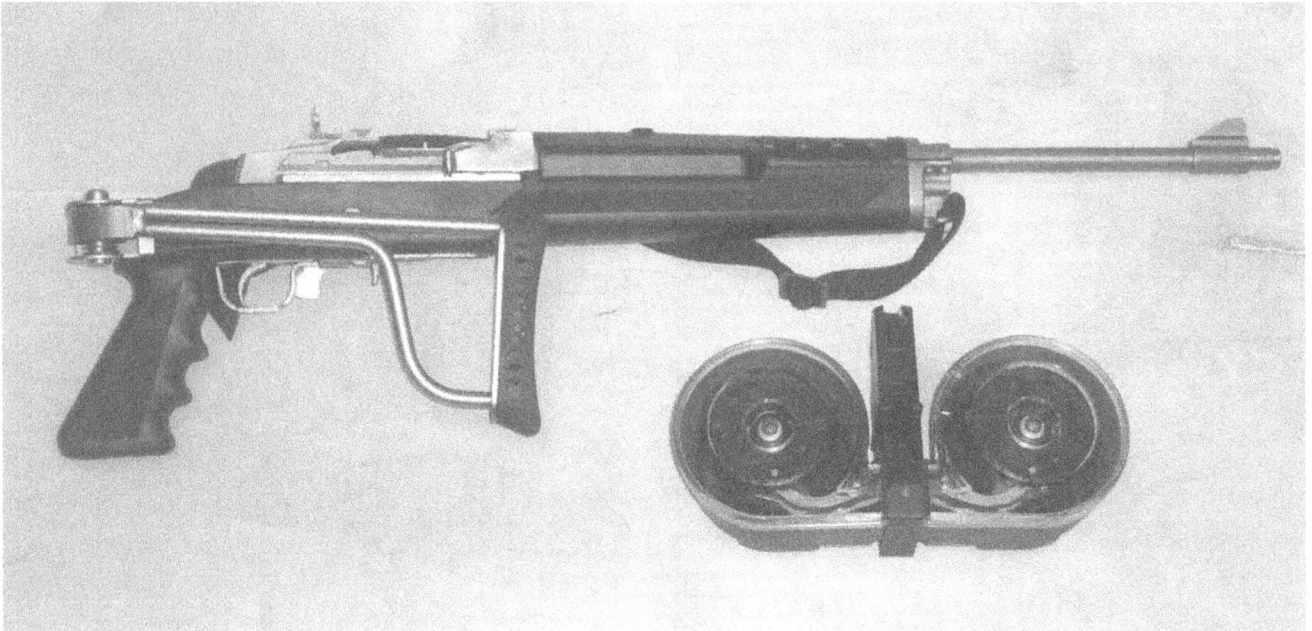

This Ruger Mini-14 carbine is fitted with a handy folding stock and will feed from this 100-round dual snail drum magazine, creating a highly capable survival tool for the apocalypse.

Accessory rails added to survival weapons for mounting sights and handles as desired.

Adjustable shoulder stock on an AR rifle.

CREATING A SURVIVAL KIT CARBINE

Some time ago I purchased a handy lever-action carbine in .44 Magnum, mainly to take along on camping excursions. I chose the .44 Magnum for this gun because the ammunition would be interchangeable with my usual camp revolvers, and also because I consider it an excellent close- to medium-range cartridge. The whole idea was for a lightweight, compact, fairly inexpensive big-bore repeater that could serve to defend against aggressive bears around camp as well as to potentially drop a deer at up to a hundred yards in any unplanned survival emergency.

The gun I chose for this was the Winchester Model 94 Trapper with 16-inch barrel that was still being manufactured at the time (around 2002) and was selling then for around $300 retail. I decided to modify this little carbine in various ways to turn it into a kind of general-purpose wilderness survival kit.

One of my favorite modifications to a wooden rifle or carbine stock is to first remove the butt plate or recoil pad from the stock, simply by backing out the two plate screws, and then drill one or two 5/8-inch diameter (or slightly larger) holes into the center of the butt (being careful to avoid removing too much material that could compromise the stock's strength). This creates a little hidden vault for any assortment of tiny survival gear items, such as wooden matches, fish hooks, sewing needles, water purification tablets, scalpel blades, bore brushes, spare gun screws or a spare firing pin, or maybe even just an extra loaded cartridge for an emergency. If a loaded round is to be thusly stored, pack some cotton in the hole with it to prevent it from rattling around inside.

A candle could be melted to drip wax onto the inside surface of the butt plate before screwing it back onto the rifle's butt, especially around the outer edge, to provide a better waterproof seal for whatever goodies are to be stored in the tiny vault. This seems like a

A hole has been drilled into the carbine butt on the left to create a small storage vault. A similar cleaning rod compartment already exists in this old military rifle's butt on the right.

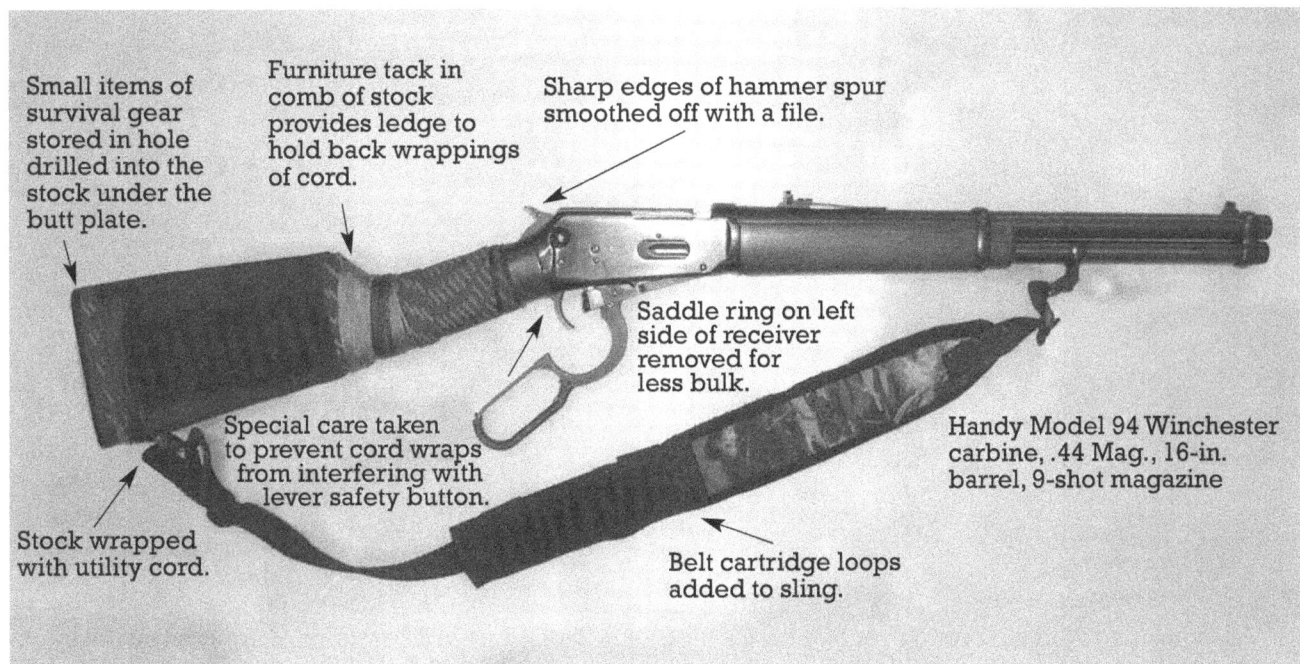

Small items of survival gear stored in hole drilled into the stock under the butt plate.

Furniture tack in comb of stock provides ledge to hold back wrappings of cord.

Sharp edges of hammer spur smoothed off with a file.

Special care taken to prevent cord wraps from interfering with lever safety button.

Saddle ring on left side of receiver removed for less bulk.

Stock wrapped with utility cord.

Belt cartridge loops added to sling.

Handy Model 94 Winchester carbine, .44 Mag., 16-in. barrel, 9-shot magazine

The Survival Kit Carbine.

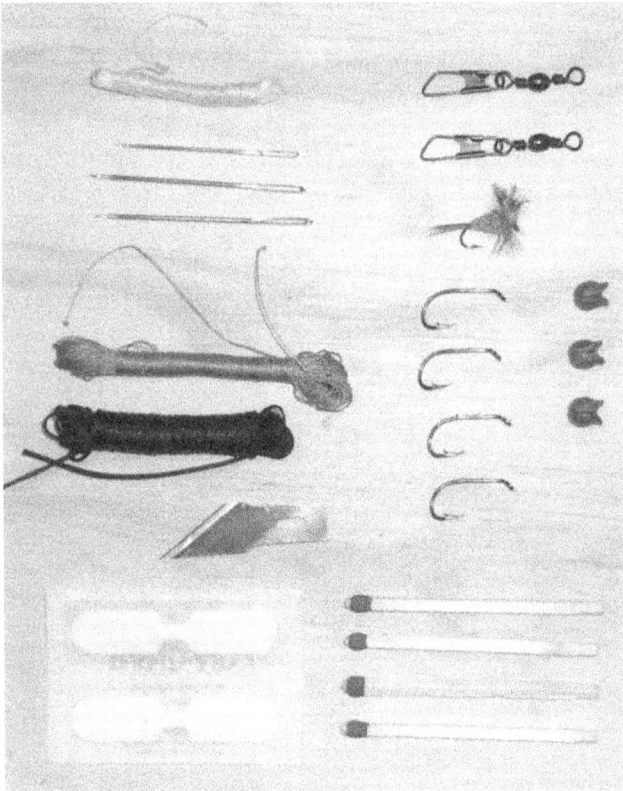

Tiny survival gear like dental floss, thread, cord, sewing needles, X-Acto blade, butterfly sutures, matches, and fishing swivels, lures, hooks, and split-shot sinkers could all be stored in a small cavity in the butt of a stock.

Note the hidden cavity that exists inside this synthetic shotgun stock. Plenty of room for a lot of small survival gear.

A key-shaped Phillips screwdriver that fits the butt plate screws is conveniently stored in this leather pocket sewn onto this carbine's sling to ensure it is always with the weapon.

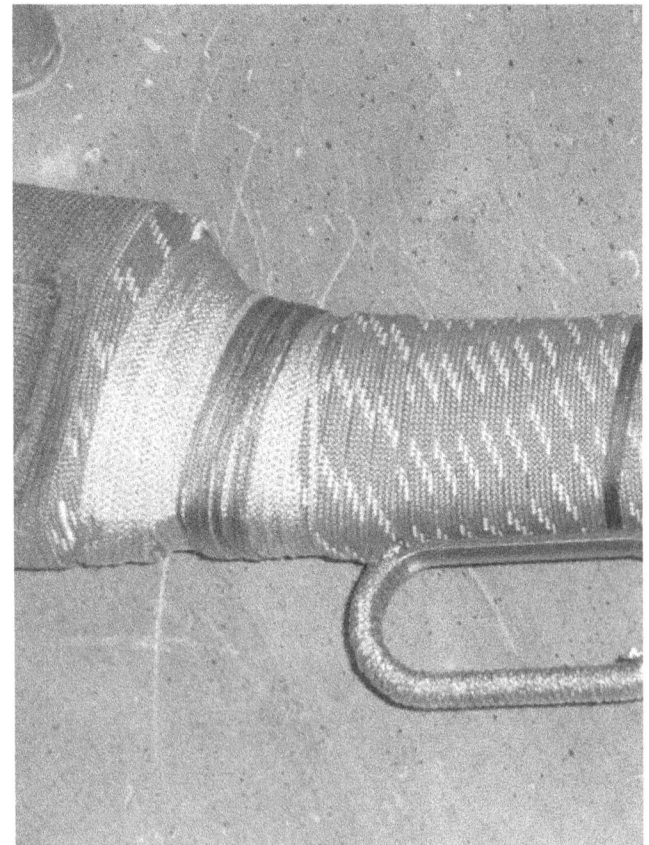

Stock and wrist portions of the survival lever-action carbine wrapped with miscellaneous utility cord.

Inexpensive Mosin Nagant bolt-action rifle before being camouflaged with camo duct tape.

more practical waterproofing method than, say, any type of glue or rubber caulking that could make the plate difficult to remove later.

I keep a small screwdriver that fits the screws holding the butt plate with the weapon at all times. I sewed a small leather pocket directly onto the outside of the sling on my carbine to securely house the little screwdriver.

Many older military rifles actually have such a hole in their butt, covered with a trapdoor for quick access, to house the cleaning rod for the weapon. I have used those handy compartments to store other small items besides cleaning kits. Also know that many modern synthetic gunstocks are actually hollow inside, and these can provide a surprising amount of volume capacity for storing survival gear.

Another adaptation I made to my handy carbine for camping and survival, as can be seen in the photos, was to tightly wrap the stock with various sizes of cordage. My idea was that if I ever happened to have that particular weapon with me in a survival situation, I would also automatically have a supply of cord for snare traps, nets, bowstrings, fishing lines, clotheslines, hammocks, extra boot laces, tent ropes, frame lashings, and so on.

The tricky part about wrapping the stock of a rifle with cord is that its shape changes; the widest part at the butt end tapers down to the wrist or pistol grip area. Keeping windings of cord tight together over something with a gradually tapering shape like that can be a challenge if you progress in the wrong direction, especially over the smooth surface of finished wood. A thin layer of duct tape adhering to the stock

under the cord can help prevent the windings from sliding to some degree, but I also noticed that if I started the cord wrapping at the smaller diameter end, in the pistol grip area, and continued up the widening part toward the butt, the tendency for the wraps to slide apart was effectively prevented.

Whenever wrapping a gun stock or other item with cord, the end result is longer lasting and better looking if the wraps are tight and close together. Also, I avoided wrapping the barrels of any of my firearms because barrels do tend to get hot after a lengthy shooting session, and synthetic cord melts easily. This would not be a concern with a rifle chambered for .22 rimfire, of course, as its barrel will probably never get that hot.

Wooden rifle stocks are very traditional and, depending on the quality of wood and finish type, can be aesthetically pleasing to a lot of shooters. However, synthetic stocks are more impervious to wet weather conditions and possibly more durable overall. For this reason, I replaced the original wooden stock on my civilian M14 rifle with a tough nylon stock, and I believe this makes it more suitable to the harsh conditions of the coming apocalypse, should something like that occur in my lifetime.

If camouflage is desired for a survival weapon, spray paint might be the way to go. Durable paints of various greens, browns, grays, and other earth-tone colors, especially matte no-shine types, can serve this purpose well. Take care to first remove or thoroughly mask off any parts you wouldn't want painted, such as scope lenses and adjustment rings, action bolts, lubrication points, or any moving parts.

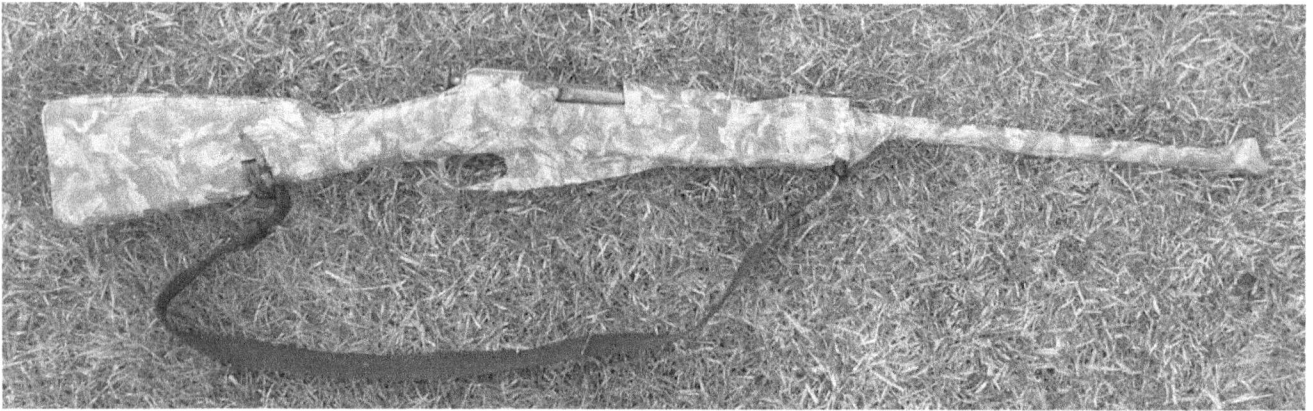

The same rifle after being almost completely taped up except for the sling.

A possible downside to camo painting a firearm, especially if the task is hastily executed or if an inferior quality paint is used, is that the paint could gum up things and interfere with the smooth operation of the weapon. Nevertheless, this camouflaging method worked to my own satisfaction on two of my long guns.

Nowadays you can find camouflage-patterned duct tape, and this opens up another viable possibility for a quickie (though very temporary) camo job that could make your firearm less visible to your enemy or prey in the wild.

I did notice with my own experiments using the duct tape camouflage method that the sticky tape tends to peel off bits of old varnish from a stock when it is eventually removed, and it appeared to adversely affect the condition of the bluing on the barrel as well. So I would certainly think twice about using duct tape to cover over any rifle having a nice finish on its metal or wood.

HANDGUN GRIPS

One of the first things often replaced on a new handgun is the grip. This is especially true with revolvers, and particularly the older revolvers that were commonly sold with small grip slabs that didn't fit a lot of shooters' hands completely or properly, at least not for optimum shooting comfort.

The problem with a lot of the older stock revolver grips was that they tended to allow the second finger of the shooting hand to wrap the handle *behind* the trigger guard, where the knuckle invariably gets hammered by the back of the trigger guard during recoil. This was more of a problem with the large-frame, big-bore, double-action revolvers than with the smaller framed guns or the traditional single-action grip designs, which tend

Stock wooden revolver grips like those on the left are often replaced with soft rubber designs like the Pachmyer at upper right or the Hogue Grip below it.

to roll the barrel upward during recoil as opposed to slamming the gun straight back into the web of the shooting hand, as is more common with powerful double-action revolvers.

Out of all the brands of aftermarket handgun grips that have been popular over the years (e.g., Eagle Grips, Pachmyer), I prefer the ones from Hogue Grips. The company's soft rubber grips are the most comfortable in my experience, and this sentiment seems to be shared by the majority of shooters I have talked with. The Hogue design does an excellent job of filling that space behind the trigger guard on the double-action guns, keeping that second finger knuckle clear of the trigger guard's hammering.

RIFLE SLINGS

We considered rifle slings in the previous chapter as weapon support components, but slings also have several nice possibilities for modification as well. A lot of interesting and useful gear can be secured to a rifle sling, ensuring that the items will always accompany the weapon.

We can attach a variety of little pouches or pockets for storing things like cleaning kits, pocket sewing kits, tweezers, small knives, takedown/disassembly tools, button compasses, bore lights, extra firing pins or other small spare parts for the weapon, or, as already discussed, a compact screwdriver to remove the butt plate.

It might be of some practical value to punch a series of evenly spaced holes along both edges of a plain leather rifle sling and lace parachute cord through the holes in order to ensure you will always have some handy small-diameter rope with you when hunting with that particular rifle.

I also found that a thin folding saw, lock-blade knife, gunsmith screwdriver, or other tool of similar size could be sandwiched between the two folded-over straps of leather at the base of a thick cowhide rifle sling, and then wrapped with utility cord to hold it in place.

Cartridge holders/loops, as anyone who's watched movie Westerns knows well enough, were traditionally made of leather and usually sewn directly onto holster belt rigs or worn as bandoleers. As much as I have

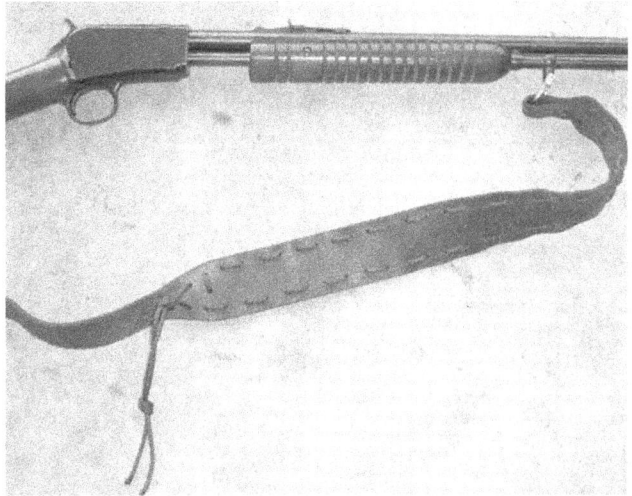

Parachute cord is laced through the holes as a way to carry a supply of emergency survival cord with the gun.

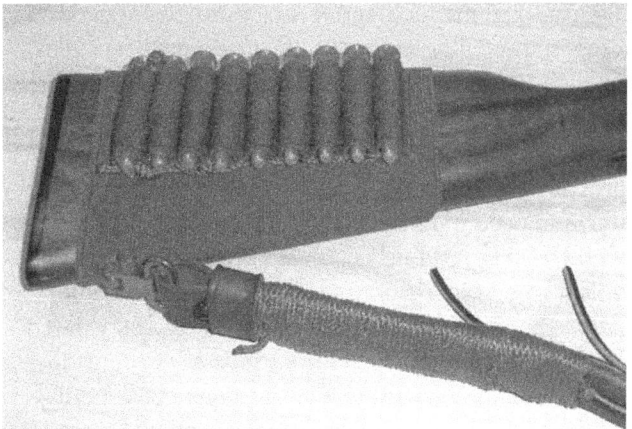

The folded double base of this rifle sling is wrapped with utility cord to secure a narrow tool between the straps.

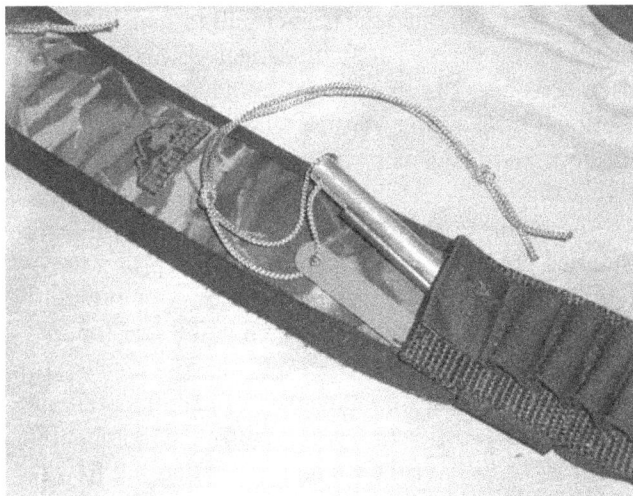

A survival fire-making spark tool—a magnesium bar with a ferrocerium rod embedded in it—stows securely inside a belt slider shell holder that has been stitched to this rifle sling. The hacksaw blade makes a perfect scraper to generate sparks and light fires with the ferrocerium rod.

Punching the holes in a plain leather rifle sling.

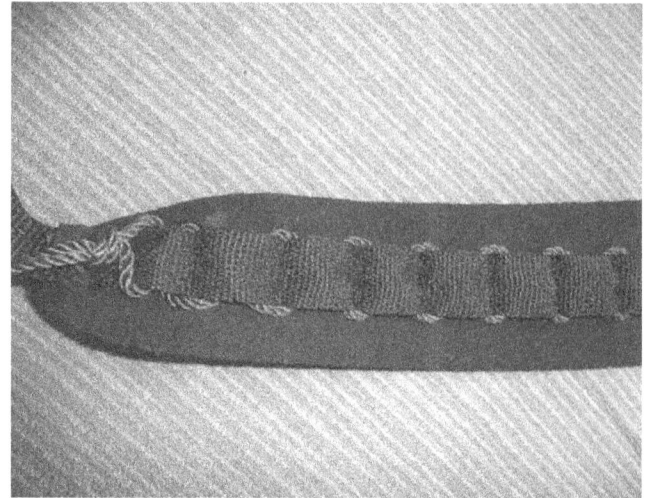

Parachute cord laced through loose cartridge loops will help hold cartridges firmly in place.

always liked the look and feel of genuine cowhide, keep in mind that leather products require more routine maintenance than do most manmade materials to keep them in service, typically benefiting from the application of saddle soap, neatsfoot oil, or some other preservative to keep the leather from drying out, cracking, and becoming hard and brittle. Also, everyone who has ever left brass cartridge cases in a leather bandoleer or belt loops for an extended period of time knows about the gummy and sometimes crusty green crud that accumulates on the surface of the cartridges.

Nowadays, cartridge loops are more commonly composed of nylon cloth or similar synthetic material and incorporated into what I like to call belt sliders that fit onto belts or backpack shoulder straps, or those elastic tubular shell holders that go on rifle or shotgun stocks. You can buy rifle slings with cartridge holders already sewn onto them, or you can add belt sliders to your sling and maybe sew them into the desired position the same way I did.

I have noticed that elastic shell holders do occasionally lose much of their elasticity as they age, especially with the larger-sized loops that hold shotgun shells or big-bore rifle ammunition. Sometimes they get stretched to the point that the cartridges are no longer held very firmly in the loops.

My makeshift solution to this annoying little issue, and one that has worked well for me over the years, is to weave parachute cord through the loops to provide a kind of filler, making the space inside those elastic tubes smaller. This easy modification gives them a tighter grip on the cartridges.

TRIGGER AND TRIGGER GUARD MODIFICATIONS

A fairly common modification to both revolvers and automatic handguns is lightening and smoothing their trigger pulls. Since the trigger action of any firearm is a major contributing factor to its performance with respect to accuracy, improvements in this area can make a significant difference.

Even so, this is one area of firearm modification that I normally tend to shy away from, simply because I've seen custom trigger jobs on double-action revolvers that lightened the trigger pull to such a degree that the guns seemed actually dangerous for general use. I've always been rather paranoid about the unintentional discharging of rounds from guns having hair triggers, and that was my worry with the trigger jobs I've seen.

Also, this type of modification sometimes involves thinning down the hammer spring (main spring) to lighten its tension and ease the double-action pull, in conjunction with altering the notch in the sear for an easier single-action trigger let-off. I have witnessed too many misfires that were the result of hammers failing to strike with sufficient force. I realize that competitive target shooters have their unique priorities, but for my purposes (in this case arming for the apocalypse), I want my weapons to be as safe and reliable as possible, so I am willing to adapt to the factory trigger pull, even if it is slightly on the stiff and heavy side occasionally.

For bolt-action rifles, a number of aftermarket adjustable triggers are available, the most famous

Close-up of double-action revolver's trigger face that was polished smooth.

Cutaway trigger guard of an old double-action revolver to make shooting easier with a gloved finger.

brand perhaps being Timney (timneytriggers.com). Everything I've ever heard about this type of upgrade has been positive. If I were to build a custom rifle, I would seriously consider replacing the stock trigger with one of these units.

A custom trigger feature that I don't see very often but one that seems like a good idea is the trigger stop, or over-travel stop. This will normally appear as a small button or lug mounted to the backside of the trigger or affixed to the inside of the trigger guard directly behind the trigger, where it will block any excessive movement of the trigger as it is depressed by the trigger finger. Trigger over-travel (where the trigger continues its backward movement after the hammer or firing pin begins moving) is one possible cause of shots being pulled off target.

Occasionally, the trigger face of a revolver can benefit from a simple modification. Some years after purchasing my Smith & Wesson Model 29 double-action revolver, it occurred to me that the series of deep grooves in the face of that gun's wide target trigger, which ran vertically and parallel to one another, were really only practical for single-action shooting, if even for that. They tended to grip the trigger finger at a single position rather than allowing smooth movement of

the finger over the face of the trigger, as is more natural during double-action shooting.

My dad's idea in this case turned out to be successful, and this was to use a Dremel Moto-Tool fitted with cylindrical grinding stones and sanding drums to remove the grooves and ultimately polish the face of the trigger. We didn't bother to remove the trigger from the gun for this process but instead secured the grip frame in a vise for the task and applied thick layers of tape inside the trigger guard and frame to mask and protect those surfaces we wanted to avoid grinding. This worked well, and I find the modified trigger much more suitable than it was previously with the grooves.

Handguns with custom cutaway trigger guards don't seem as common now as they once were, but occasionally one encounters an old revolver with this type of alteration. I found a Colt Army Special with the front of its guard partially cut away, and because of this single issue I was able to buy it for a fraction of the gun's unaltered value. The gun is perfectly functional in every way. The only reason I can image someone wanting to cut a trigger guard in this way is to make it potentially easier to use while wearing heavy winter gloves. This is not a modification that I would personally recommend under normal circumstances.

CHAPTER 9

Alternative Weapons to Consider for the Apocalypse

In the postapocalypse novel *Dies the Fire* by S. M. Stirling, firearm ammunition is rendered inoperable by some mysteriously unexplained event that simultaneously causes aircraft to fall out of the sky, motor vehicles to stop running, and the power grid to shut down. In that fictional nightmare, people are forced to rely on medieval swords and other edged weapons, bows and arrows, and crossbows for self-defense because firearms simply would not work.

Granted, it was just a fictional scenario and perhaps an implausible one at that, but the entertaining adventure does provoke some thought about alternative weaponry. I can certainly envision scenarios where we might not have access to our firearms when the major world event occurs. We could be traveling in a foreign country or for whatever reason find ourselves stranded many miles from home, away from our arsenal of personal survival weapons. Under circumstances completely outside of our control, we may not find local gun stores open from which to buy whatever we might need.

You might wish to think of alternative types of weapons as strictly auxiliary, secondary, backup, or "last-ditch" options to your primary first-grab weapons. Keeping them in this context will help you more clearly envision their merits for the apocalypse.

Within this alternative weapon category, we will include, among numerous non-gun weapons, the old-fashioned muzzleloading firearms and even modern air-powered guns, because these are not classed as firearms by the Bureau of Alcohol, Tobacco, Firearms, and Explosives in the United States and as such are not subject to the same restrictions. (For more information on the legality of every class of firearm, visit the ATF's website at www.atf.gov.)

FLINTLOCK FIREARMS

The flintlock as a potentially viable postapocalypse firearm system is considered here for the following reasons:

1. The weapon's low-tech engineering. A blacksmith with any gunsmith skills could repair and possibly even build from scratch a complete and operational flintlock gun using unsophisticated eighteenth-century skills and technology.
2. Its comparatively basic operating requirements. It requires black gunpowder and the sharp piece of flint for ignition.
3. Its time-proven usefulness.

The flintlock is not dependent upon metallic cartridge cases, precision-sized or copper-jacketed bullets, primers, or percussion caps of any kind. The gunpowder itself could *potentially* be homemade, with the most common recipe simply consisting of 75 parts saltpeter (potassium nitrate), 15 parts charcoal, and 10 parts sulfur by weight, finely ground separately and mixed together wet before being dried in small quantities for safety reasons. (For an in-depth study of making gunpowder, get a copy of *The Do-It-Yourself Gunpowder Cookbook* by Don McLean, published by Paladin Press.) Functional projectiles could easily be fabricated out of a variety of materials.

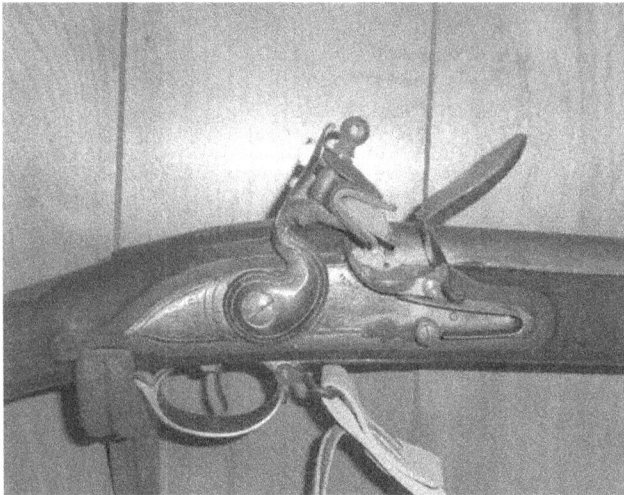

The lock of a Brown Bess flintlock musket.

Every part of this functional lock was made by Gene Ballou from a single truck spring. Now if only he would get busy and build the rest of the gun for it!

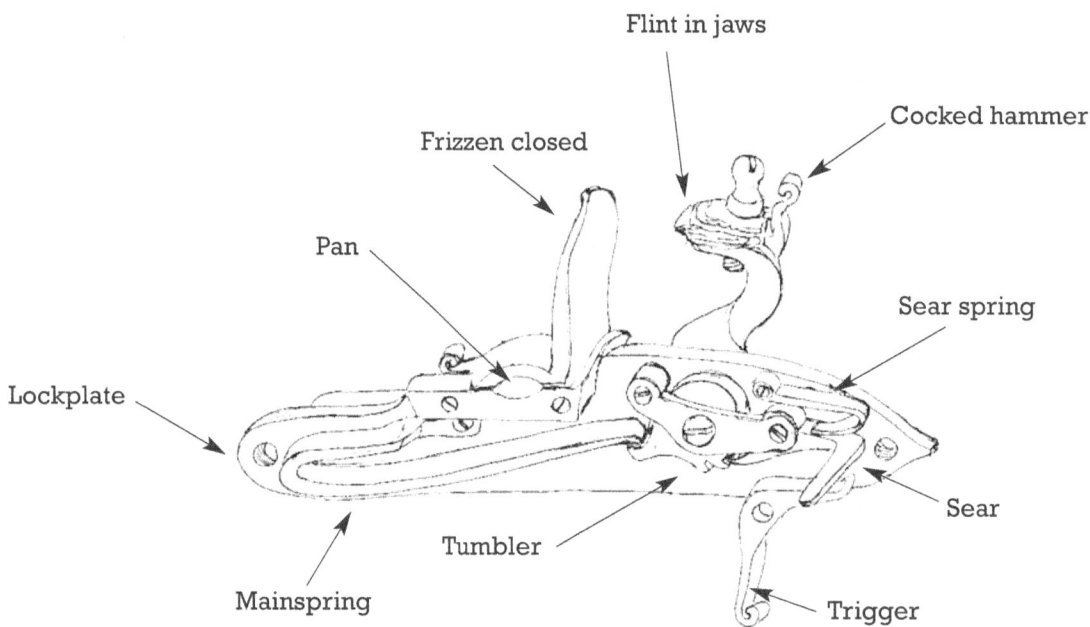

Flint in jaws

Cocked hammer

Frizzen closed

Pan

Sear spring

Lockplate

Sear

Tumbler

Mainspring

Trigger

Inside view of a flintlock mechanism.

A third reason to consider the flintlock as a serious survival weapon is its time-proven usefulness; it remained in service as the dominant firearm system from the late 1500s with the earliest "snaphaunce" flintlocks until almost 1840—roughly 250 years, as compared to the 50-year lifespan of the percussion system that replaced it, or the now almost 150 years of metallic cartridges. Our last reason for considering this type of weapon in this context is because a flintlock is a lethal tool, capable of killing animals for meat or other desperate survivors trying to kill you.

Flintlock guns were the primary arms used in

These three long guns all have flint ignition.

.735-in. diameter
lead musket ball
Mass: approx. 585 gr.

.45 ACP, loaded cartridge
Mass: approx. 335 gr.

Size/weight comparison of a musket ball and a
loaded .45 ACP round, for perspective.

numerous bloody wars fought all over the world, accounting for hundreds of thousands, if not millions, of casualties, and they helped colonists, pioneers, frontiersmen, and native tribes gather meat for their families for literally centuries, and on more than one continent. In spite of its many shortcomings (which we will examine shortly), the flintlock can be used effectively for hunting any kind of game, for recreational target sports, and for personal self-defense.

In the late 1700s and early 1800s, long guns existed as either rifles, muskets, or "fowling pieces," which were basically the earliest shotguns. Guns with rifled barrels naturally tended to be more accurate than the smooth-bored muskets, but they also demanded more care in their loading process, because bullets were normally patched with cloth for a tighter fit inside the bore and to grip the rifling grooves, and powder charges would often be more carefully measured and consistent. Bore sizes ranged widely, but calibers from about .36 up to about .60 inch were the most common in the long rifles.

By contrast, the muskets were usually larger weapons with larger bore diameters, being intended for close-quarters volley fire. The idea was to rain as much lead into the enemy ranks as possible, with little or no regard for the accuracy of individual weapons. Musket bore diameters ranged from .69 to about .80 inch. The English Brown Bess muskets most commonly featured .75-caliber, smooth-bore barrels. They were often fired with buck and ball loads (one large ball plus several smaller balls of buckshot) fitting loosely in the barrels, typically housed in cloth cartridges together with their powder charges for the fastest reloads. In the reproduction version of this weapon that I own, a .735-inch round ball will fit firmly in the barrel using T-shirt cotton or bed linen material for the patch. A lead ball of this size weighs considerably more than an ounce.

Although a gun with a smooth bore will certainly not be as accurate as one with rifling, it will be easier to clean, faster to load and reload, and not as particular in its loading requirements. Having such a large bore size, the musket lends itself well to functioning as a shotgun.

A wide variety of projectiles have been fired out of smooth-bored guns, from patched balls and bullets of every configuration to bird shot and buckshot, copper-plated BBs, steel ball bearings, air gun pellets, darts, creek gravel, brass tacks, split-shot fish sinkers, broken glass, marbles, and rock salt. While much of this sort of makeshift fodder cannot be recommended here, I think it is useful to note the versatile nature of the smooth-bore musket. I have loaded my musket as a shotgun on several occasions with good results, using felt wads intended for loading 12-gauge brass shotgun shells, since the bore size is roughly the same.

We also have to weigh the drawbacks of flint ignition, especially by today's standards in order to be as objective as possible about their utility in an apocalypse situation.

Flintlocks (as well as all muzzleloaders, for that matter) are slow to load and reload by modern-day standards. The most common configuration is a single-shot, muzzleloaded affair, not particularly ideal for a firefight where the enemy may be armed with automatic weapons or at the very least repeating cartridge firearms. Even with premeasured charges of powder in capsules for quick reloads, with loading blocks full of patched balls at the ready, and maybe an extra ramrod or two within easy reach, the front stuffer simply cannot compete with modern firearms when it comes to rapid fire.

Since the flintlock operates using black powder, each shot produces copious amounts of fire and smoke. More than just the inconvenience of temporarily blocking the shooter's sight picture, the visual signature of the smoky explosion could potentially give away his position to enemy observation.

And then we have the reliability issue to contend with. Simply put, flint ignition is not, never was, and never will be nearly as reliable as either the percussion cap that replaced it (after the cap was eventually refined) or especially the metallic cartridge that is commonplace in our time. When a sharp edge of flint strikes briskly against carbon steel, sparks *usually* fly.

That's about the only way I know how to correctly describe a process that is as primitive as the rock-against-steel ignition system. When conditions are just right—and they do normally have to be just right—with the lock parts clean and having good, strong mainspring and frizzen springs; the flint of correct dimensions for the lock size and having a sharp, even edge firmly secured in the jaws; the frizzen properly tempered; the touch hole clear of obstructions; and the powder in the pan of the finest grain and thoroughly dry, the system *usually* works! Even when sparks do actually shower nicely down from the frizzen of the lock as intended, at least one of those sparks must reach the priming powder in the pan, and that powder must be bone dry for there to be quick ignition. Misfires and delays in ignition are not uncommon in rainy or damp weather.

The cleaning and routine maintenance requirements of any flintlock weapon are properly described as "demanding" when compared to modern firearms using smokeless ammunition. If the bore of a black powder gun is not swabbed and cleaned routinely—ideally between every two or three shots fired through a rifled bore; somewhat less frequently with a smooth bore—the buildup of fouling will soon impair the continued

Wyatt Rogers is engulfed in smoke after he takes a shot with a .69-caliber flintlock smoothbore.

The dramatic flash in the pan of a flintlock musket. To make it easier to capture the ignition for the photo, the black priming powder was mixed with modern smokeless rifle powder, but smokeless gunpowder should, of course, never be used in any gun barrel made for black powder only.

performance of the weapon. Also, if black powder guns aren't cleaned well within a reasonable amount of time after use, the residue could present a corrosion problem on the surface of the gun's metal.

There are modern black powder substitutes that are supposed to be cleaner burning, but my understanding is that, at least the most popular of them, Pyrodex, isn't as sensitive a propellant as true black powder and really wouldn't be very practical for use in guns with flint ignition. I stick with traditional black powder in my muzzleloaders.

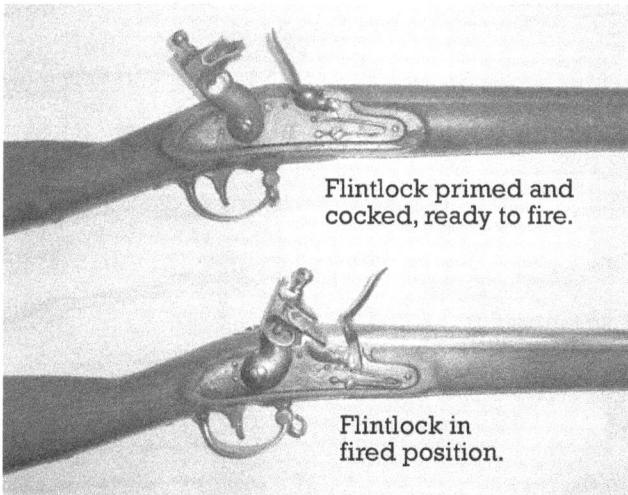

Flintlock primed and cocked, ready to fire.

Flintlock in fired position.

Flintlock muskets in both ready-to-fire and un-fired positions.

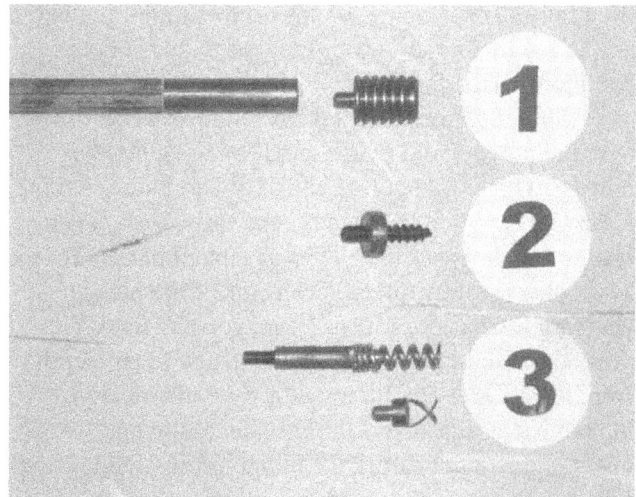

Commonly used muzzleloader ramrod tips: 1) ribbed cleaning jag, 2) ball puller screw, and 3) two variations of the patch puller worm. All will thread into the end of this ramrod.

Black gunpowder is available in several "Fg" grades of fineness. FFFFg is very fine and used mostly in the priming pans of flintlocks. FFFg is used mostly in .32- to .45-caliber muzzleloader barrels. FFg is popular in muzzleloaders .50 caliber and larger. Fg would be suitable for cannons with 1-inch or larger bore sizes.

Additionally, traditional black powder is a versatile type of gunpowder in the sense that it can be used in muzzleloaders as discussed here, as well as loaded in numerous metallic cartridges for modern handguns, rifles, and shotguns (just as it was before about 1895). It has been used in cannons, fuses, fireworks, rockets, and explosives for centuries. If forced to grab only one container of gunpowder while bugging out for the apocalypse, I would simply go for the genuine black gunpowder, because with it I could probably make almost any of my guns work. The tightest six-shot

group I ever obtained on paper with my Smith & Wesson .44 Magnum measured 1¼ inch center-to-center and was well centered in the target, and I did it off-hand from 42 feet using my experimental hand loads consisting of .240-grain, lubed, hard cast lead semiwadcutter bullets over 30 grains of FFFg GOEX black powder. My nickel-plated revolver was surprisingly easy to clean afterward, using a stiff nylon brush and warm water with dish soap.

Also, for a general-purpose gunpowder, I would opt for a courser grade of powder suitable for large-bore muskets (GOEX FFg), because the larger granules can be ground finer to serve other purposes in a pinch.

One final drawback to any muzzleloader is the fact that a loaded weapon's charge in the barrel cannot quickly and conveniently be unloaded without discharging the weapon.

A common mistake—and I've done this many times—is to forget to pour the powder in before ramming the patched ball firmly down the barrel. This is often the result of daydreaming, talking with a buddy, or guzzling too much beer during the loading process. It was also a common error made by soldiers amidst the chaos of eighteenth-century battles.

In such cases, a shooter can, without having to take the barrel out of the stock and the breech plug out of the barrel, attach a woodscrew-shaped tip to the end of his ramrod and turn it with hand pressure against the front of the ball until it bites and threads into the soft lead, and then pull the bullet out. Sometimes that pointed screw tip threaded onto the ramrod can be used for pulling stuck cleaning patches out of the barrel, although another device called a worm is more commonly used for that.

Another common mistake is to ram the ball down over the powder and fail to seat it fully against the powder. This is sometimes called a "short seat," and it is easy to do when the bore is especially dirty. It can lead to bulging, cracking, or even a burst barrel. To avoid short seating, and to determine if the powder was forgotten as described above, you can mark the ramrod to facilitate a quick visual gauge of the depth from the muzzle end for a specific powder charge. If the ramrod doesn't reach all the way to the mark when inserted down the barrel, the bullet is not seated properly or you forgot to include the powder.

The problem of rain and moisture reaching the powder in the pan was always a concern with flintlocks. There are several possible remedies for this situation. Perhaps the highest quality modern lock on the market that I am aware of is the famous Chambers Round-Faced English lock from Jim Chambers Flintlocks (www.flintlocks.com). Not only is this a majestically styled flintlock with a sturdy precision mechanism that throws plenty of hot sparks, but it also has a unique raised waterproof pan that will minimize the moisture problem. The Chambers lock is more expensive than some of them, but it is truly excellent.

Old-time trappers and mountain men often covered the locks of their guns with a rain bib, typically just a piece of deer hide that had been treated with tallow or grease to help it shed water. It was fitted over the lock, where it could be tied down to keep the gun's action covered and dry while the frontiersman carried his gun out in the rain. This bib could be quickly untied at one end and pulled back out of the way to allow the lock to function whenever a shot needed to be taken.

Modern flintlock guns are available from a variety of sources and within a wide price range. This eliminates any reason to put more miles on any surviving antique guns from the period, whose values depend so heavily on their condition.

Home workshop rifle building became a popular hobby with black powder shooters decades ago, when a multitude of kits and component parts became widely available, and many of those products are still available. We can buy virtually any of the period-styled but newly made parts and build our own guns.

Unlike my dad, I am no master craftsman or hobby gunsmith by any measure, but for an experiment, I set about to assemble a makeshift flintlock

The excellent Round-Faced English lock with raised waterproof pan from Jim Chambers, here with the frizzen open to show the pan.

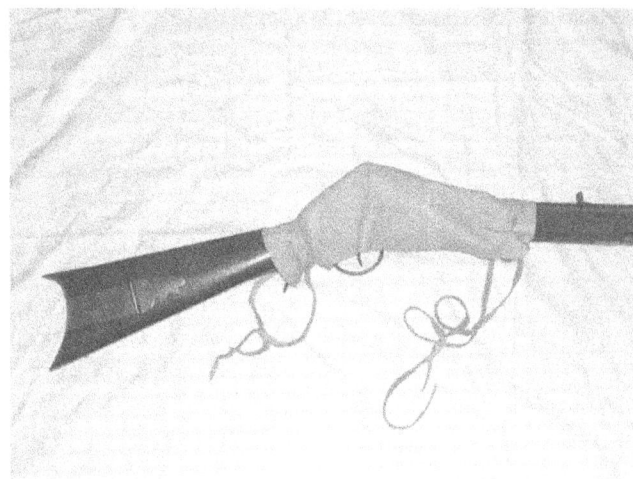

Buckskin rain bib tied down over the gun lock.

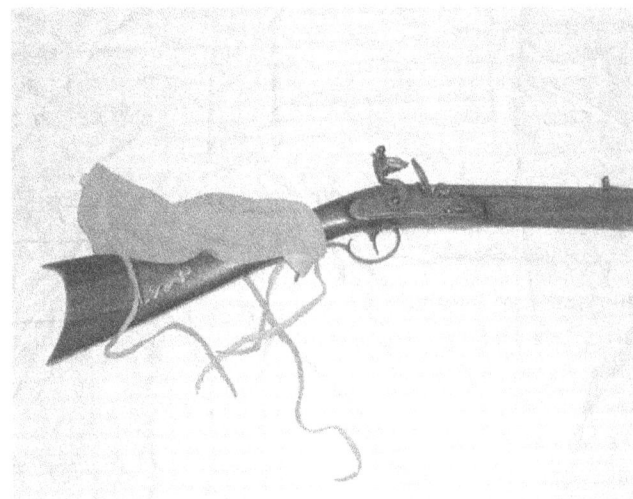

Rain bib flipped back out of the way to make the gun ready to shoot.

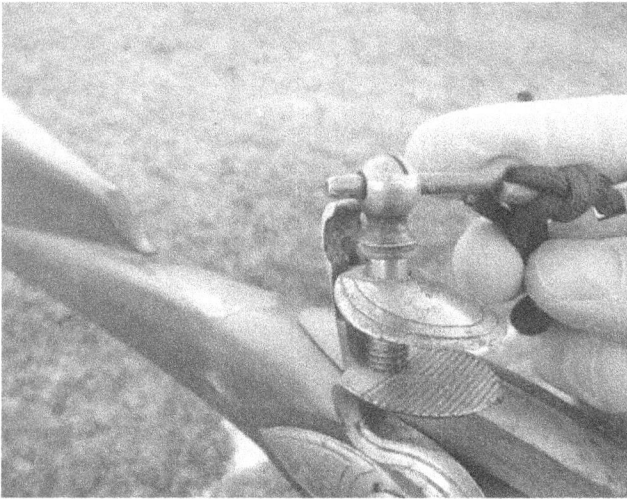

The jaws of a flintlock can be opened with a screwdriver or sturdy nail, punch, or steel rod if the top jaw screw has a hole through it, as shown here.

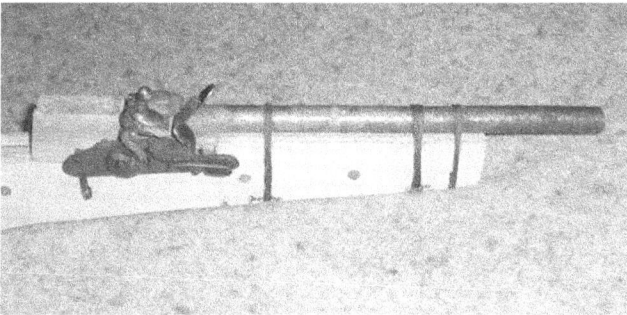

A makeshift muzzleloader pipe gun that uses flint ignition.

The flintlock pipe gun and everything needed to load and fire a shot: FFFg black powder for the main charge, ramrod/cleaning rod, lead ball, cloth patch for the ball, FFFFg powder in a priming horn to fill the pan, and an adjustable powder measure (adjusted for 20 grains in this case).

weapon to see just how achievable such a task might be for the average guy who may need to improvise some type of effective close-range weapon in a postapocalypse emergency.

I happened to already have a small, inexpensive, imported lock to start with—I think it originally cost around $25—but I had no barrel, trigger, or other components. I decided to improvise those items from common household hardware. A section of ½-inch galvanized steel plumbing pipe would serve as the barrel, and the breech end was closed with a steel end cap and copious amounts of solder to build up that area. (I believe a much safer chamber could be devised using a steel coupling over the pipe's threaded end in conjunction with a solid pipe plug to close the end.)

The crude stock was assembled from three layers of half-inch-thick plywood screwed together with sheetrock screws, and the lock was screwed to the plywood on the right side of the breech where the lock's pan aligned with the vent. I drilled the vent (i.e., touchhole) into the chamber area using a #56 drill bit and attached a small strip of sheet metal to the sear such that it bent around to the face of the lock in order to act as a kind of side trigger lever, which actually works quite well. The barrel, as can be seen in the photos (at left), was secured to the makeshift plywood stock with bindings of strong braided cord.

This expedient flintlock carbine is completely functional, such as it is, and I test fired it using a .437-inch diameter, 124-grain lead round ball, cotton cloth patch, and 20 grains of FFFg GOEX black powder. I was not eager to increase the powder charge beyond 20 grains with the relatively thin pipe barrel.

I discovered that at point-blank range, even with the light 20-grain charge (and a certain percentage of the propellant always blows out of the touchhole of a flintlock), the ball easily shot through a 1-inch pine board. In my experiment, the ball also made a noticeable impression in the log positioned behind the board.

This degree of penetration may not seem too impressive until one attempts the same feat (as I did) using the same size lead ball with a high-powered wrist-rocket slingshot, drawing the pouch back to 3½ feet for the maximum hitting power. The slingshot-propelled lead ball, which I suspect could be deadly if a target were hit in the head by it, merely dents the face of the board.

I should warn readers that fabricating any type of expedient pipe gun, even one intended for a relatively light charge of black gunpowder as described here, is a

Front side of pine board shot with the makeshift flintlock pipe gun.

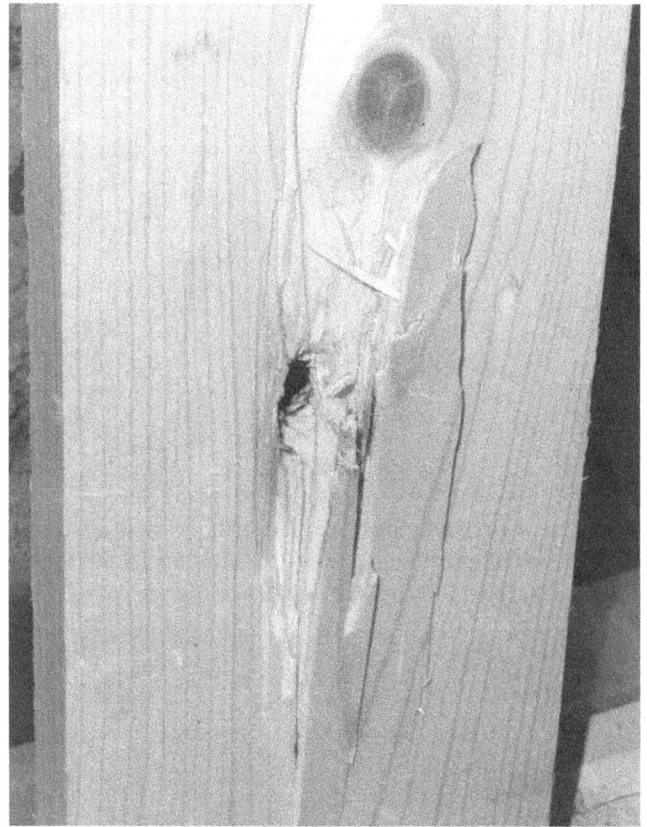

Backside of pine board, showing exit hole of the ball.

risky endeavor. Just because my experiment did not happen to blow up on me does not mean that yours won't. I only describe it in this context of *last-resort, field-expedient* weapons that we might devise in a postapocalypse emergency, when the desperate need for anything that shoots outweighs the risks.

Readers interested in purchasing newly made flintlock weapons, kits, and reproduction antique parts (as well as other period-related gear) may wish to contact Dixie Gun Works (www.dixiegunworks.com) or, for quality, period-style smoothbore guns and kits, North Star West (www.northstarwest.com). Conveniently for the modern shooter, there are a number of historical reenactors and early lifestyle enthusiasts who keep the period gun fascination alive. Just about anyone wishing to add a flintlock or two and the related accoutrements to their long-term survival arsenal can easily do so thanks to the ongoing interests of this "buckskinner" culture.

To operate any muzzleloader, a portable kit containing the necessary equipment will be needed. The frontiersmen often carried, in addition to their powder horns, what was commonly called a "possibles bag," which

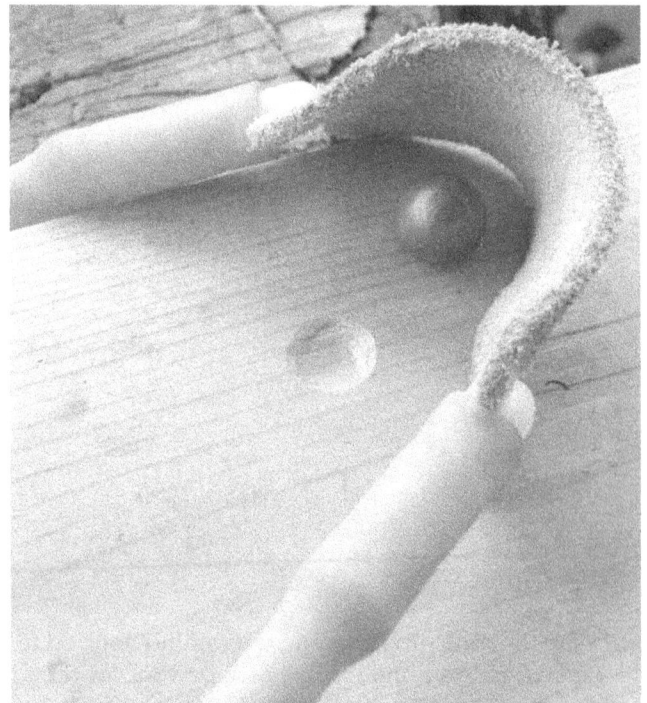

The 1-inch pine board is only dented by a .44-caliber lead ball launched with a high-powered slingshot at a distance of 2 feet.

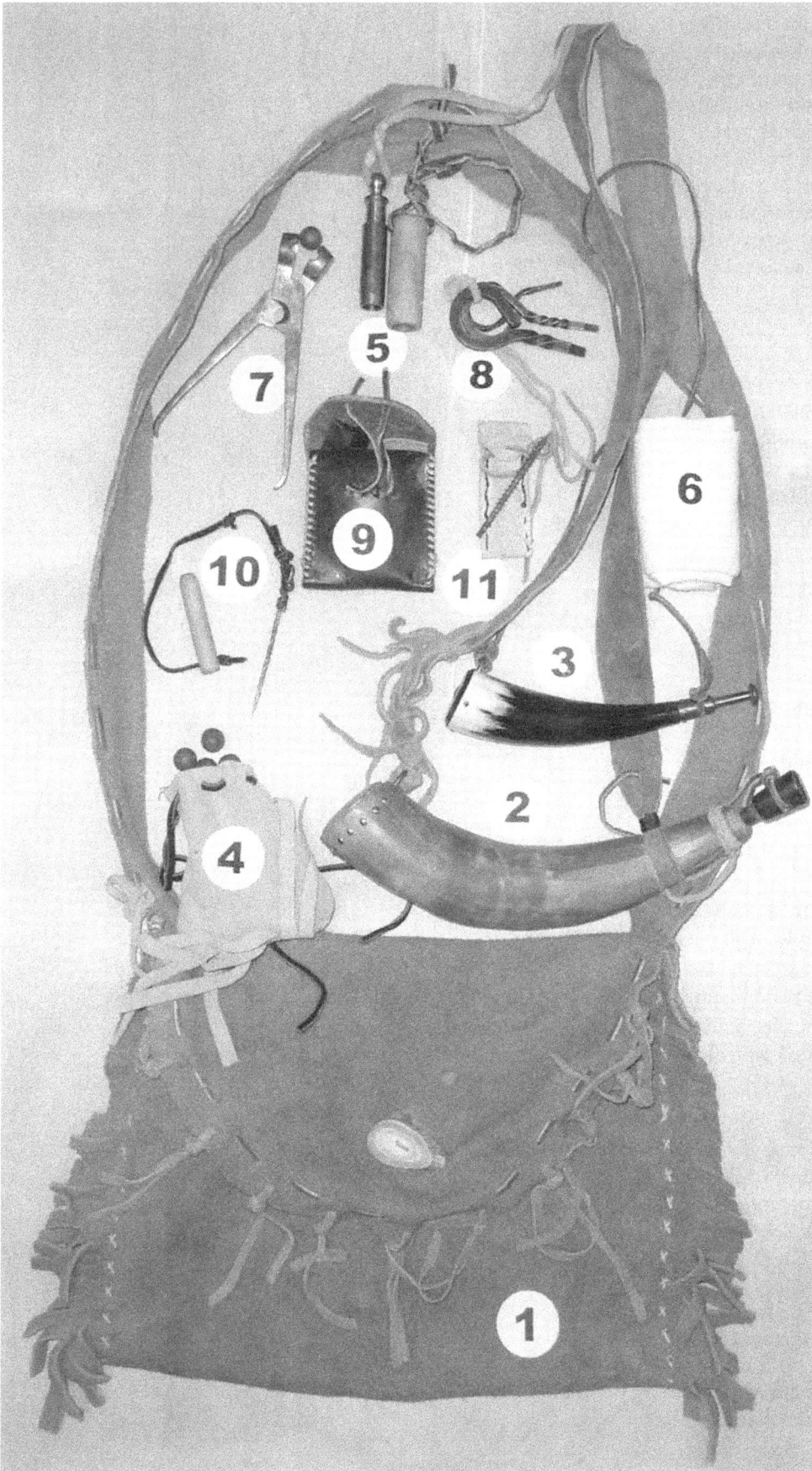

The basic shooting kit of the flintlock shooter: (1) elk hide bag to house the paraphernalia, (2) large powder horn holding the gun's main powder charge, (3) smaller priming horn holding finer powder for the lock's pan, (4) lead round ball bullets in leather bag, (5) adjustable brass powder measure, plus a wooden fixed powder measure that holds 85 grains of FFg powder, (6) cloth rags for ball and bore patches and general gun cleaning, (7) small brass bullet mold, (8) screwdrivers that fit the screws of the gun's lock, (9) small leather pouch containing extra gun flints and flint jaw pads, (10) touch hole clean-out pick and its point cover, (11) bullet puller screw fits end of ramrod, with its own protective rawhide pouch.

A few more tools that might find utility in the flintlock shooter's kit: 1) straight razor as an efficient patch-cutting knife—the attached slotted thong slides firmly over the closed tool to prevent the blade from opening in the kit, 2) bullet starter/short ball rammer, 3) small pliers useful for knapping/restoring sharp edges of gun flints, and 4) mainspring vise for lock assembly/disassembly in the field.

was really just a shoulder bag to contain most of the necessary accoutrements for their weapons, plus maybe some small survival items like flint and steel for making fires.

It may seem like the shooter who uses a flintlock sure needs a lot of tools and gear to shoot and maintain his weapon, and this cannot be denied. But we have to remember that he is in fact hand loading every shot he makes in the field, and that process always requires certain tools and components. I think it's useful to note that many of the flintlock's basic accessories, like powder flasks/horns, bullet blocks, powder measures, and touch hole picks, could be (and often were in the old days) manufactured by the shooter. This is usually not the case with the individual who uses cartridge-firing modern firearms exclusively.

Survivors of the apocalypse may also find portable quick reloads handy in an emergency. Premeasured powder charges housed in individual capsules, smaller priming powder flasks for convenient pan refills, and a bullet block that holds a series of pre-patched

At top we see how these one-shot quick reloads work. The large capsules hold the main charge of powder, while the smaller ones hold finer powder for quickly priming the pan. The capsules are simply pulled free of their stoppers, which are secured to a belt or gear bag, thus making them instantly ready to pour. They can be used in conjunction with the bullet block (at bottom) holding pre-patched balls at the ready for quickly reloading a muzzleloader.

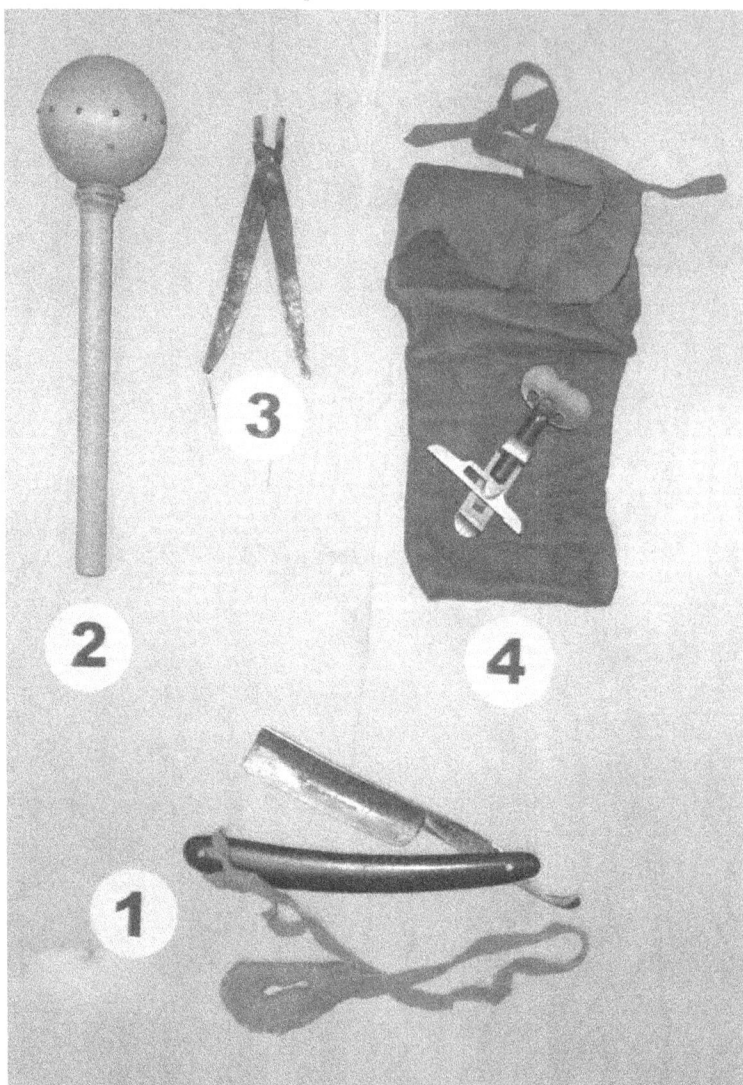

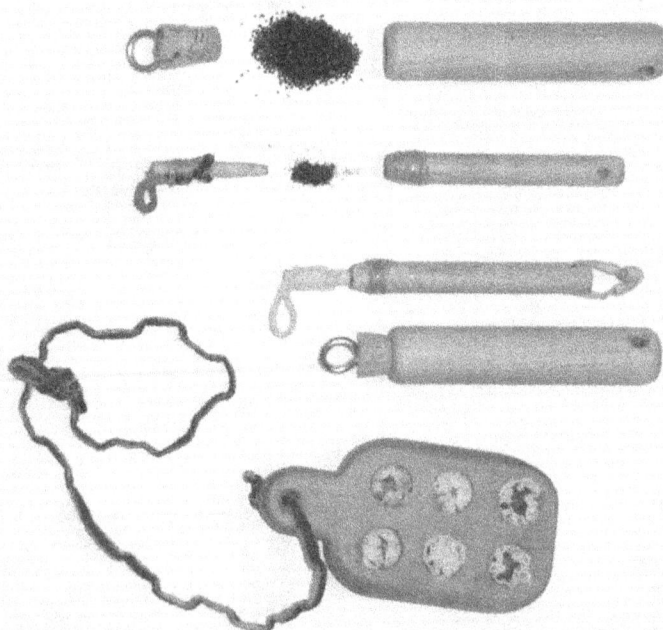

Using a bullet starter to load a musket from a bullet block.

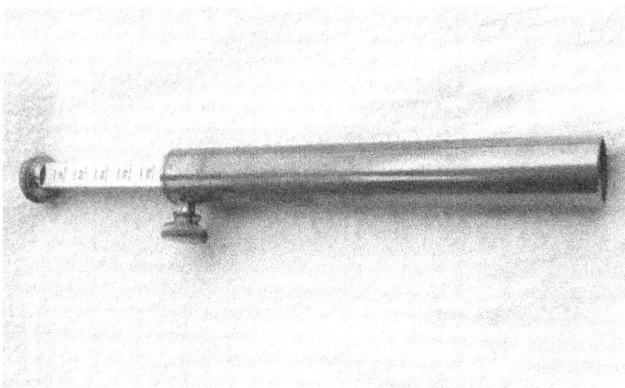

An adjustable powder measure that can measure up to 120 grains by volume, indexed in 5-grain intervals.

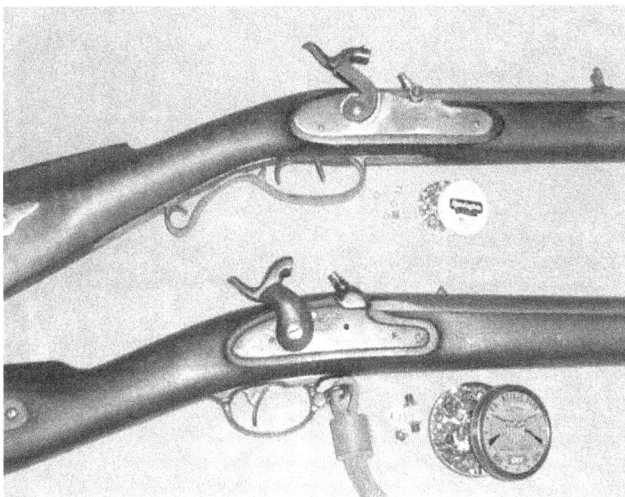

The two common cap-lock sizes: the plains rifle at top uses #11 caps, and the Model 1841 Mississippi rifle below it uses the larger musket "top hat" caps.

balls at the ready can save the shooter a lot of time with his reloads—and maybe even save his life!

I found it easy to make these quick-reload capsules by drilling a hole into one end of a wooden dowel large enough to contain the desired amount of gunpowder. I capped the openings of the main charge capsules with corks that I inserted eye screws into so they could be hung from a belt loop or any strap, whereby a moderate tug on the body of the capsule, using only one free hand, uncorks it and readies it for pouring. These smaller priming flasks/capsules have their own tapered, soft wooden stoppers with similar loops attached for this same readiness.

A word of caution is warranted here, given the sensitive nature of black powder. The capsules containing powder should be kept safely clear of flames, sparks, or the explosions in the flash pan during flintlock ignition.

Similarly, it is recommended that a shooter never pour powder from the powder horn or flask directly into the barrel of a muzzleloader but instead first pour into a powder measure (or use the quick loads described above) and then dump the measured charge down the barrel. Not only could a tiny ember down in the chamber, possibly still glowing from the previous shot, ignite the powder and follow it up to the horn or flask in the shooter's hand, but whatever powder charge is poured down the barrel should be measured for obvious safety reasons.

The balls in the bullet block are held in place by their precut cloth patches. Patches lubed with bacon grease or Crisco will tend to give better performance—they will not burn up as quickly when fired and will produce less fouling in the muzzleloader barrel than will plain dry patches.

Firearms that use the percussion system will be more dependent upon factories than will flintlocks, given their requirement for percussion caps. However, they are also more reliable and tend to give quicker ignition than flintlocks. An apocalypse survivor could have several thousand percussion caps saved and stored within the space of a small coffee can, and that amount might be made to last for years.

There are two common percussion cap size categories: the #10 or #11 caps that are standard with cap-and-ball sporting rifles, shotguns, and pistols, and the larger musket-type "top hat" caps for use with the Civil War-era muskets and rifle-muskets.

Aside from the priming component, muzzleloaders with the percussion cap-lock system load and clean exactly the same as flintlock muzzleloaders.

PNEUMATIC (AIR) GUNS

Just as is true with the loose powder-and-ball type guns, air rifles and spring-powered BB or pellet guns do not fit the contemporary legal definition of "firearm" in this country and therefore are not subject to all of the same restrictions, at least not with the federal gun laws. This is a point worth considering by modern-day preppers. (Still, it is your responsibility to research and comply with all local, state, and federal laws before buying or building any firearm.)

Also worth contemplating is the economy of a gun type that uses air or the power of a spring to propel small-caliber projectiles. Most air rifles cost less than a new cartridge firearm, and a few ammo cartons that each contain several hundred pellets collectively costing maybe $10 or less could last a survivor quite a long time.

It is easy to view this whole category of weapon as hardly more than kids' toys, but as we shall see, some of them have much more potential than that. Air rifles capable of propelling lead pellets at 500 to 1,000 fps (and in some cases even a few hundred feet per second faster) will be useful for pest control and harvesting birds and small mammals for food in both rural and urban environments. They would be economical for target practice, and, in the hands of skilled marksmen and under special circumstances, they *could* potentially serve as effective close-range defense tools.

Clearly, not even the highest velocity air rifles

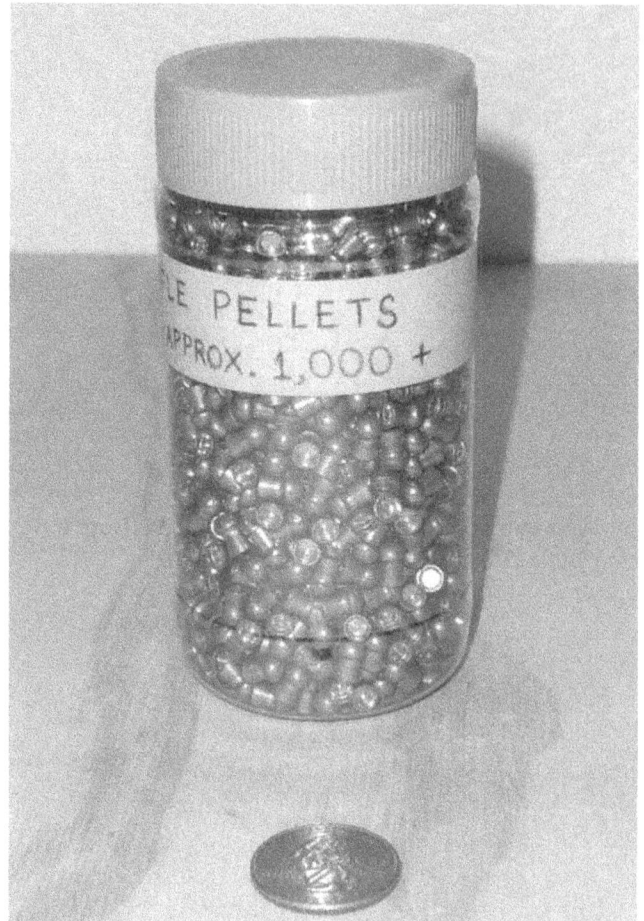

This little plastic jar, shown close to a quarter to underscore its compact size, holds more than 1,000 of the .20-caliber lead pellets.

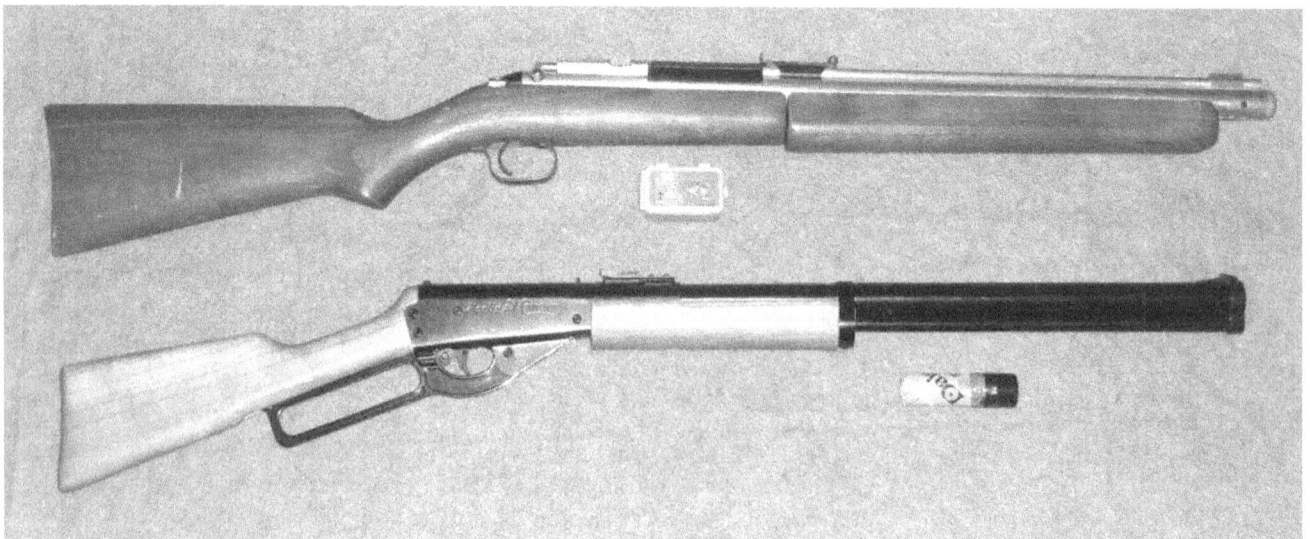

The Sheridan air rifle pictured at top currently retails for under $200 and shoots 14.3-grain .20-caliber lead pellets from a rifled barrel at around 650 fps with eight to ten pumps. The Marlin/Crosman lever-action spring-powered BB gun at bottom retails at anywhere from $29.95 to $45 and shoots copperplated steel BBs at around 300 fps from a smooth-bore barrel.

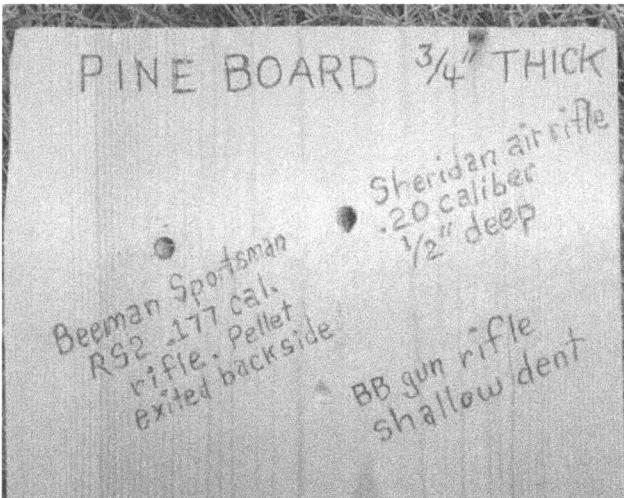

Pellet penetration test into a pine board at point-blank range. The Sheridan air rifle penetrated more than halfway into the board with 10 pumps, while the Beeman break-action rifle drove its pellet clear through. My son's 300 fps BB gun merely dented the front side.

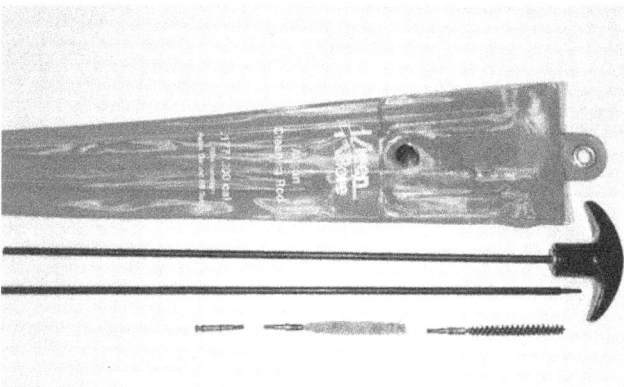

Kleen Bore's Airgun Cleaning Rod with several different tips is the right size for pellet guns.

This Benjamin/Sheridan pump-action pellet pistol shoots 5mm (.20-caliber) lead pellets. The owner of this gun told me he has killed large birds with it.

offer exceptional knock-down power compared with any cartridge firearm. Pellets in calibers .177, .20, or .22 inch are typically available in weights from close to 6.0 grains on the lighter side to about 18 grains on the heavier side, with a few exceptions that fall slightly outside this general range. But if we use a relatively high average weight of, say, 12 grains to the pellet, and assume a relatively high velocity of 1,000 fps, then using that formula we learned earlier (multiplying the grain weight of the projectile by the velocity squared and then dividing that by 450,436), we discover that our "high-powered" air rifle delivers only a mere 26.64 ft-lbs. of muzzle energy! By contrast, nearly every .22 LR bullet available today will generate at least 100 ft-lbs. from a rifle barrel, and most will give around 130 ft-lbs. of muzzle energy. Nevertheless, a high-velocity air rifle will easily bury a pellet deep into a plywood or pine board at close range.

In my own experience with air rifles, I have found their cleaning requirements to be minimal compared with any other type of gun, because they aren't plagued with the fouling of hot gases or burnt powder residue. This quality would be a distinct advantage for any long-term survival weapon all by itself. However, a cleaning rod with bore cleaning attachments specifically made for .17- and .20-caliber bores will come in handy for lead removal in the bore, like the little kit made by Kleen Bore (www.kleen-bore.com).

I believe it is best to avoid pellet guns that are powered by CO_2 cartridges or any other cartridge/canister that contains air or gas under pressure. Not only would their resupply become unreliable after the apocalypse, but they are bulky, expensive, limited in purpose, and don't last as long as they should.

Besides copperplated steel BBs that weigh barely more than 5 grains apiece, air guns having rifled barrels will normally shoot either .18- (.177-inch diameter) or .22-caliber lead pellets. The only other caliber that is somewhat common in the small-bore air rifle category is the .20-caliber (5mm) pellet used in Sheridan air guns. Given the typical weight range for air gun pellets noted above, this is what I would call true economy of weight/size in ammunition.

While a $30 Daisy or Crosman BB gun might be considered perfectly suitable as a youth's first rifle, possibly even adequate for knocking sparrows off a telephone wire from 30 feet, and perhaps ideal (and a lot of fun) for indoor target practice due to its low noise and very limited power, some of the harder-hitting air rifles available today are some serious weapons.

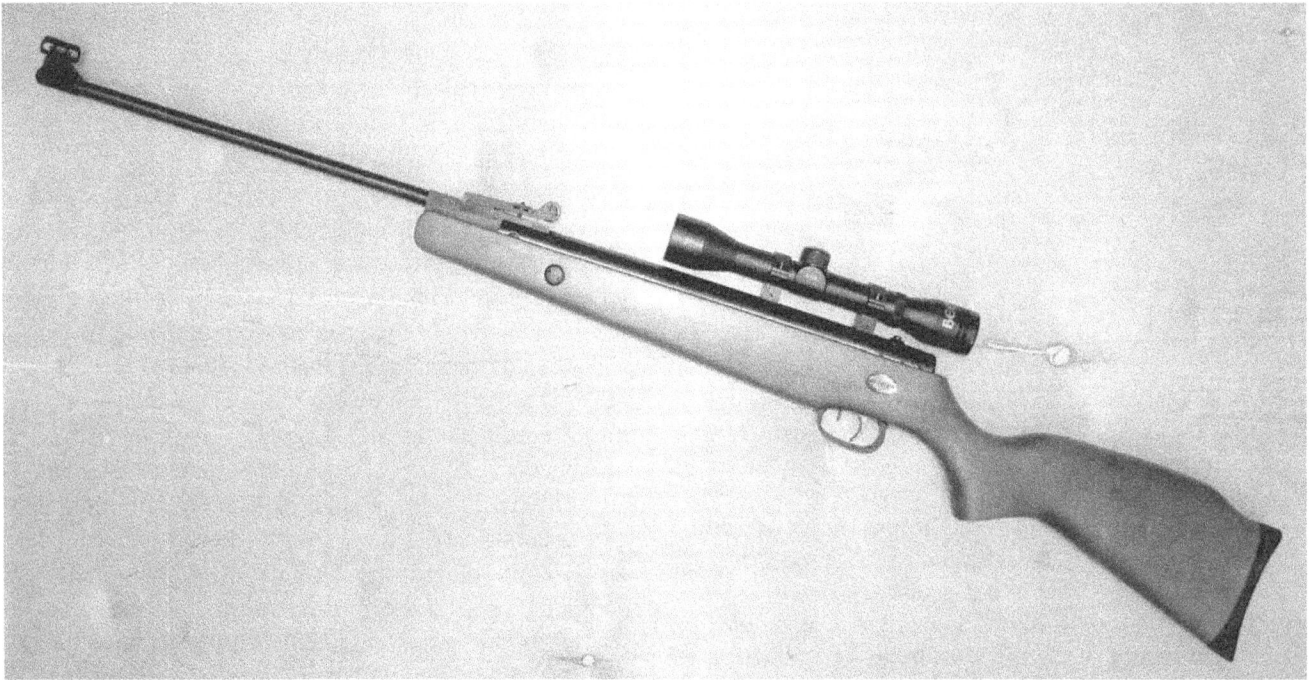

The Beeman Sportsman RS2 air rifle with 4x scope.

Beeman's Sportsman RS2 Series, for example, is a break-action, spring-powered, .177-caliber air rifle with a rifled 18-inch barrel, equipped with fully adjustable open sights and an optional 4x scope, capable of firing a pellet at 1,000 fps muzzle velocity, if we are to believe the manufacturer's specifications on the box. Other powerful spring-piston (as well as gas spring) air rifles are available from GAMO, Stoeger, Theoben, Weihrauch, Crosman, Benjamin, Umarex (Ruger), and others.

One of the great things about the break-action, spring-powered air rifles and pistols so popular nowadays is that you only have to cock them once to make them ready to fire, as opposed to the traditional multistroke air guns that require a series of pumps (usually five to ten, depending on the desired power), in addition to cocking their trigger springs/bolts to prepare them for the shot. Progressively more muscle is required to pack all that air into their diaphragms, especially during the final pumps. Additionally, a spring compressed with one stroke will provide more consistent (and potentially higher) velocity than will the system that depends on a variable number of pumps to compress air.

Nevertheless, any air-compressing pump-action or multistroke air rifle, such as those made by Benjamin or Sheridan, will deliver respectable (for an air gun)

A 7.9-grain, .177-caliber, flat-nosed lead air gun pellet from Daisy next to a .22 Long Rifle cartridge to show size comparison between these two small rounds.

Cocking the power spring of the Beeman Sportsman RS2.

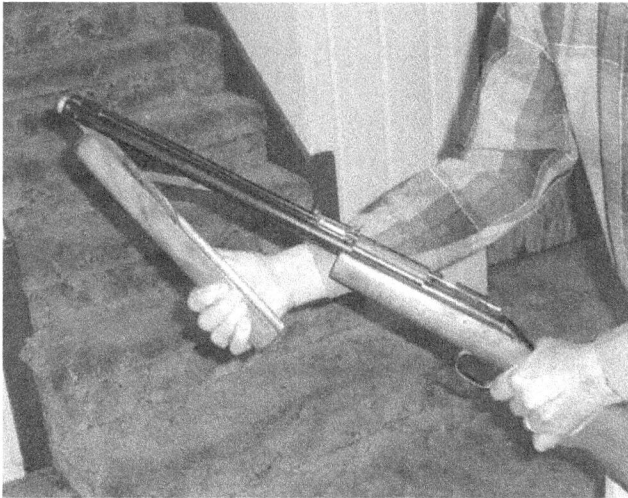
Pumping the handle of a Sheridan air rifle to compress the air.

The Sheridan has a bolt that must also be cocked to the rear before the gun will shoot.

power and accuracy, fully capable of killing pigeons, chickens, lizards, lake turtles, squirrels, and other small critters out to about 20 yards.

ARROW AND DART LAUNCHERS

If the loud sound of discharging firearms makes them impractical under special circumstances, or if ammunition resupply becomes an issue, the go-to weapon of choice often will be some type of bow, crossbow, blowgun, slingshot, or other primitive projectile-throwing device. Within this realm, the ancient bow and arrow is still a viable and popular weapon technology in our modern era, certainly for hunting and sporting purposes, if not so much as a combat weapon.

It is important to understand that the penetration characteristics of a typical hunting arrow set it apart from a firearm bullet. Being of comparatively heavy mass and having a sharply pointed broadhead tip, even flying at a much slower velocity than most handgun or rifle bullets, an arrow tends to slice and glide its way through certain target materials that would easily stop bullets discharged from small arms. According to the Wikipedia article on ballistic vests, "It is common knowledge that an arrow broadhead, such as those used for hunting, will pierce all known fabric vests designed to stop the blunt impacts of bullets."

The velocity of hunting arrows launched from contemporary bows normally ranges from around 225 fps from a very fast recurve bow to well over 300 fps from the fastest compound bows of today. But even a lightweight arrow that leaves the newest compound

bow's rest at a sizzling 340 fps travels significantly slower than the slowest centerfire round in existence.

I divide the general field of modern archery into four subcategories. Let's have a look at each.

Compound Bows

Nowadays, we have the popular compound bows with their pulleys, cables, peep sights, sophisticated trigger releases, carbon arrows with plastic vanes, and all sorts of technological advancements. Look in an issue of *Field & Stream* or any other mainstream hunting magazine and you'll see plenty of examples of modern compound bows and their paraphernalia, because they are what most contemporary bow hunters seem to be using now.

For situations that call for precision accuracy with arrows, especially at the outermost limits of effective arrow range, it would be hard to beat a serious bow hunter who is highly skilled with a state-of-the-art compound bow. My former neighbor, Matt Kelso, happens to be one of those. While most of the bow hunters I've talked with over the years seem to agree that 40 yards is about the outer edge of practical bow hunting range, I have witnessed Matt group his arrows into his target cube with amazing consistency beyond 50 yards and even from a rooftop at a range of 68 yards!

The compound bow uses a system of pulleys (called "cams") mounted on its limbs that provide the shooter enhanced leverage for a mechanical advantage in the "let off" stage of the draw, where he can more comfortably hold his arrow for aiming with less effort at full draw.

Bow hunter Matt Kelso aiming his Hoyt Carbon Element compound bow with Carbon Blade stabilizer, Black Gold Vengeance fiber-optic five-pin sight, and Fuse drop-away.

Unlike wooden bows, a compound bow is not affected by fluctuations in temperature and humidity, and it launches arrows with much more accuracy, velocity, and distance. It is a sophisticated device that will provide unmatched performance, but not without a few trade-offs.

For one thing, compound bows require a bow press for changing strings. Most compound bow shooters must take their bows to pro shops for string changes and routine adjustments, and in a survival situation this will not be feasible. Also, a device that relies on a more complex mechanical system (as compared to simple stick bow) will also have more opportunity for mechanical failure. Finally, the compound bow uses specialized arrows composed of special materials and falling within a certain weight range that are not suitable for use in most recurve or long bows, being too light. Likewise, heavier wooden arrows suitable for traditional bows aren't considered safe for use in compound bows, because they would be subject to breaking or shattering under the greater forces of the compound bow, posing a potential safety hazard.

Traditional Archery

Our second category of archery is what is often called "traditional archery." Here is where you will find the wooden and laminated recurve bows that became popular decades ago, the English longbows of medieval times, wooden (typically cedar) arrow shafts, real bird feather arrow fletching, and twisted Dacron B-50 bowstrings (in days gone by, strings were of natural fiber, often linen). This category is what your grandfather would envision if you brought up the subject of archery.

I come upon used recurve bows and occasionally quivers full of well-used traditional arrows at garage sales around town every year, and the prices are usually very reasonable. Sometimes the used arrows will have one or two feather vanes missing (an easy repair), or the points or nocks might need to be replaced, but more often than not such issues can be corrected and the arrows once again put into service.

I have purchased a number of perfectly serviceable garage sale bows at $10 apiece, and on a few occasions for even less. From my perspective that is a good bargain, and the garage sale variety of archery tackle provides great recreation during camp outings with family and friends, without a bank-breaking investment. This caste of archery tackle could certainly serve survivors well in a postapocalypse world.

If you decide to go this route in your preparations for the apocalypse, I would simply recommend looking the gear over carefully before buying, and watch out for things like hard-to-see hairline cracks in a bow

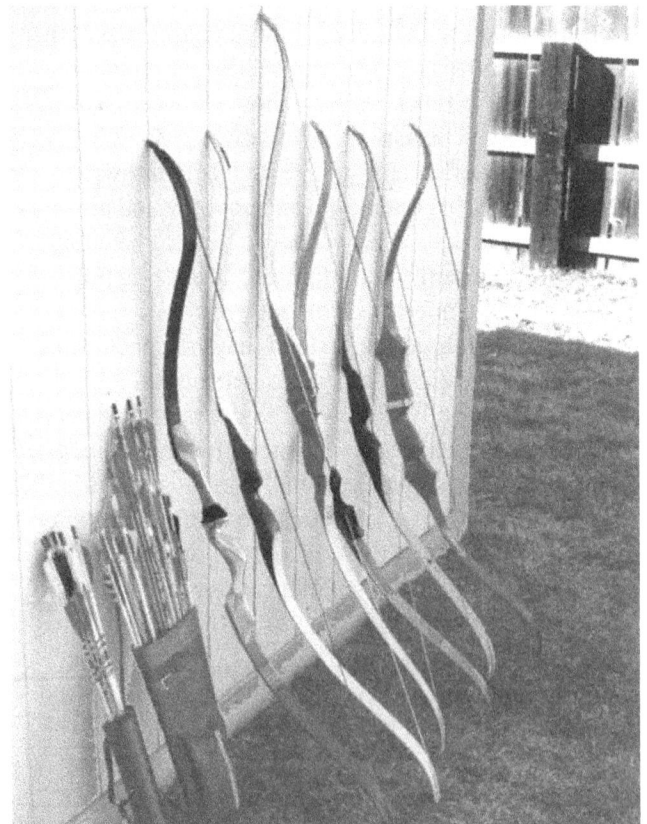

These traditional bows and arrows were purchased at garage sales over the years. They all shoot.

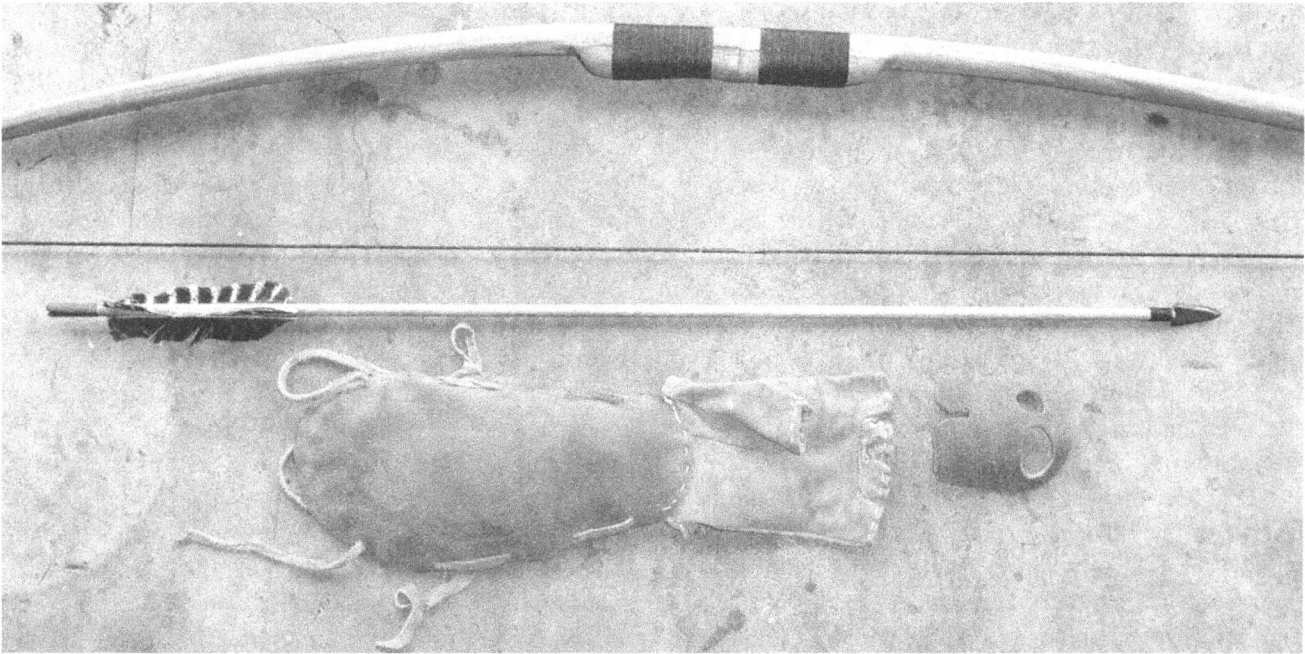

Homemade traditional archery tackle, including oak longbow, traditional arrow, cowhide arm-guard attached to fingerless glove, and cowhide shooting finger tab.

that would render it beyond repair, severely warped limbs, or laminated sections separating in subtle areas where the old glue is no longer holding things together. Some of these flaws can be difficult, if not impossible to correct. Sometimes there is a hidden reason why a seemingly quality product is set out on the driveway wearing a surprisingly low price sticker.

It is usually not too difficult to custom build quality traditional archery equipment at home, using only common hand tools and a few raw materials from the local hardware store. I built an excellent English-style longbow from a 1 x 2-inch by 6-foot-long oak board that I purchased at Home Depot several years ago. I evenly tapered both limbs using only a wood rasp and carved the functional nocks at the ends to accept the string, which I built up out of waxed Dacron cord that I purchased from an archery tackle supplier.

Arrow shafts are ideally made from the straightest hardware store wooden dowels having sufficient length. Feathers from turkeys, pheasants, ducks, geese, crows, or just about any large bird will serve as fletching material. If you can't find enough feathers in your backyard, you can buy them at craft shops or online. I also made my own arm guard and shooting finger tab from scraps of leather I scrounged up. I attached my homemade arm guard to a leather fingerless glove with stitches.

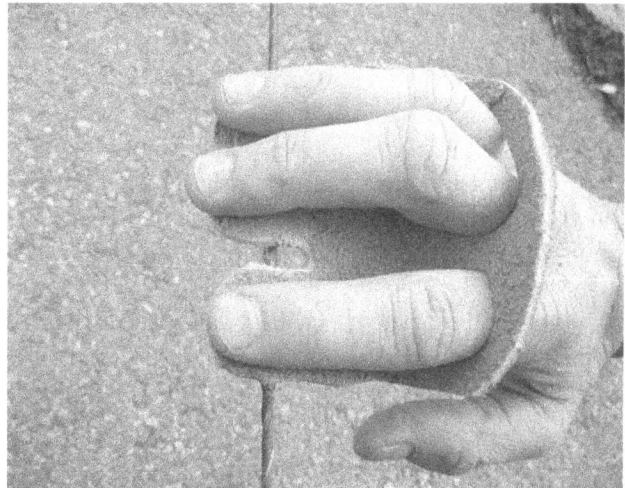

How the homemade leather finger tab works, close up.

Demonstrating how the homemade traditional archery tackle works.

Primitive Archery

The third category to consider is what has become known as "primitive archery." It includes the bow and arrow styles and technologies employed by the early American Indians, early Chinese, various aboriginal tribes, and other primitive peoples around the world. This class of archery is where we find the simplest wooden stick bows, usually either flat self bows (i.e., bows made from a single piece of wood) or sinew-backed bows handcrafted from natural materials using mostly primitive tools; strings typically made from animal sinews; arrows shaped from reeds or straightened tree branches; and arrow points usually of knapped stone or hand-forged iron.

Primitive archery could be accurately characterized as crude when compared with modern compound bow equipment. While certainly effective and deadly within their range, primitive (or even modern makeshift) bows and arrows are simply not as efficient in their function and performance as the latest marvels of engineering.

All three of the categories mentioned thus far should find a place in a postapocalypse world for those specialty applications already noted, but this third category might possibly be the most applicable how-to *skill* for survivors of an apocalypse. In other words, you should know how to build the equipment in a primitive environment as much as how to use it, simply because when it comes to resupply of bowstrings, arrows, and arrow points, it will be easier to fabricate these necessary items by hand with the most basic technologies and components using natural or commonly available materials than it will be to build or acquire the more sophisticated products.

Surprisingly simple, easy-to-make "quickie" bows fit best into this handmade category, and they can be fabricated in a pinch. Even something so rudimentary as what we will explore here could be a lethal weapon at close range.

Nearly any springy stick of roughly an inch and a half in diameter at its thickest part (preferably in the center grip area of a shaped and tillered bow, but more likely at the base end of the stick) and between four and six feet in length with notches at both ends to accept the string will suffice as a functional quickie bow. Some readers might even find usable bow staves growing on the trees in their own yard. Again, any springy stick having roughly the proper dimensions, be it green *or* seasoned, could be made to work as long as it has the necessary flex.

Let's explore this idea a little further.

Building the Quickie Bow

A very expedient weapon could comprise merely an unshaped, springy green stick (with the bark left on if time is critical) lopped from a tree in your backyard to serve as your bow, with its ends strung taut by a shorter length of parachute cord or similarly strong string, used in conjunction with an unfletched, fairly straight, skinny pointed stick (possibly clipped from the same tree) notched at the back end to fit the string as the arrow. Such a crude device might be sufficient to shish kebab a snow bunny for dinner from 15 feet away or present a shockingly unpleasant surprise to an aggressor up close in the dark. Yes, it could be just that simple.

The selection of wood for the bow will affect its performance, and some varieties of wood (yew, Osage, hickory, lemon, ash, etc.) have superior qualities applicable to the mechanics of particular bow designs than other types of wood. But in reality, most kinds of wood—save for the softest varieties, like pine—could be made to work. The only criterion is that the stick must be springy for the period of time you need it.

The finest of traditional bows have nicely tapered limbs of equal size and flex and in proper balance and tiller (i.e., the limbs bend symmetrically), but these desirable characteristics are not necessary in an expedient bow that you might need in a hurry. Just remember that what you are devising here is meant to take game or save your life in a close-range emergency situation, not to help you win an Olympic archery match.

Quickie Bowstrings

As noted, even a length of parachute cord *could* serve the purpose in a pinch if that's all that is available, although it is clearly far from ideal. For one thing, parachute cord (or "paracord," a type of small, synthetic rope) is really too thick and heavy for the most efficient string performance. And even though it will be plenty strong for any quickie bow, it will gradually start to stretch during use.

Twisting natural fibers into usable general-purpose cordage is not at all difficult. I explained in detail exactly how to do it in my first two books for Paladin Press, *Long-Term Survival in the Coming Dark Age* and *Makeshift Workshop Skills for Survival and Self-Reliance*.

However, a bowstring sustains an enormous amount of shock and wear during normal use, and rarely will hand-twisted plant fiber cordage have the necessary durability for this task. The American Indians understood this and thus made their bowstrings out of twisted rawhide, sinews, deer intestines, horsehair,

A spool of waxed B-50 Dacron cord for making bowstrings.

and various other animal products much more often than out of vegetable fibers. In some regions of the ancient world, silk strands were used successfully in bowstrings.

Our options are much broader now with the various synthetic materials found in cord today. Modern bowstrings are composed of such materials as polyester (Dacron), Kevlar, polyethylene, or some combination of these synthetic fibers much more often than from any of the natural fibers, including linen. B-50 Dacron, for example, is manufactured as cord especially for building bowstrings. This is a strong synthetic cord that resists stretching better than some.

A very expedient possibility would be to use strands of waxed dental floss bundled and wrapped together into a serviceable bowstring. A durable bowstring would require at least 12 to 16 strands of floss, assuming a tensile (breaking) strength for each strand of at least 15 pounds to be on the safe side, using the four-times-the-draw-weight rule for a bowstring, and assuming that our bow will pull 45 to 60 pounds. Dental floss is normally composed mostly of nylon, and nylon stretches more than does Dacron, so this would be a very temporary emergency-type of string for sure. I mention the possibility here only because we are more likely to have a supply of waxed dental floss than a supply of B-50 Dacron or any other popular synthetic bowstring cord during an unplanned scenario.

Quickie Arrows

Arrow shafts can be improvised from almost any small diameter (5/16 to ½ inch) wooden dowel or straight, skinny stick of adequate length. Kinks in arrow shafts can be worked straight by repeated bending with a simple wrench, ideally over the heat of a campfire.

Arrow fletching can be provided in any of a number of ways. I even read an article describing how to tie pine needles onto the backs of makeshift arrow shafts. The author reported that it actually keeps arrows flying straight, although I have not yet tried that one.

Fletching stabilizes the arrow in flight by providing drag at the back of the shaft, and anything that achieves this properly will work, although clearly some materials and designs are superior to others. The American Indians normally used bird feathers to stabilize their arrows. They tied them onto the shafts, usually with sinew, in a wide variety of arrangements, depending on the unique style of each tribe.

Those who have read my book *More Makeshift Workshop Skills* (published by Paladin Press) will know exactly how to fabricate expedient fletching from duct tape, which represents just one very effective possibility. Your options are really limited only by your imagination.

We also have plenty of options when it comes to

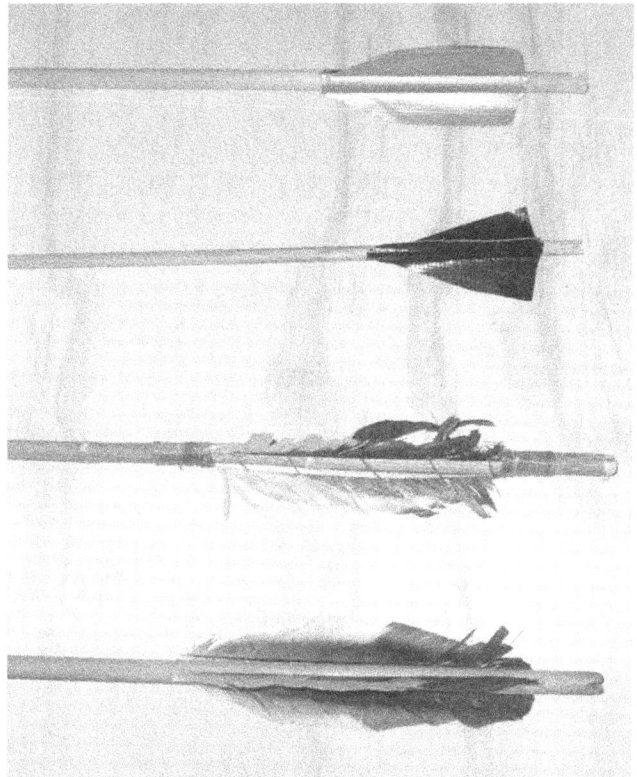

A few examples of expedient arrow fletching, top to bottom: duct tape (silver and black), seagull feathers tied onto the shaft, and feathers glued on.

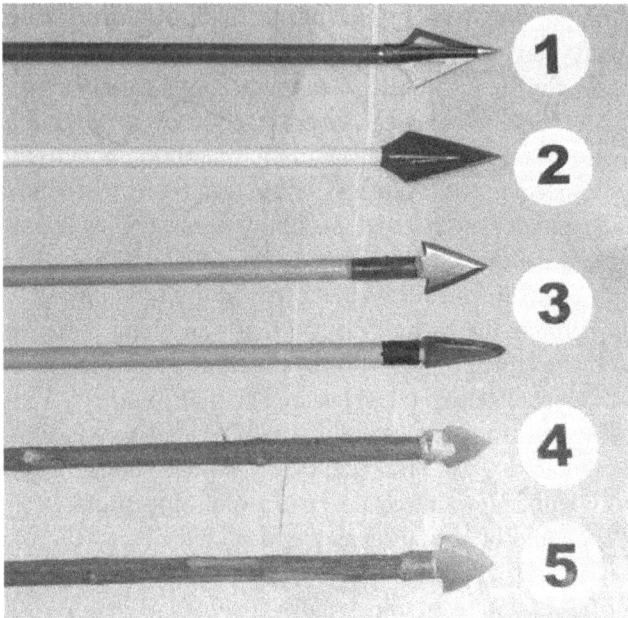

Possible arrow points for the apocalypse: 1) three-blade broadhead that screws into this carbon fiber shaft, 2) traditional steel broadhead mounted onto a cedar shaft, 3) two styles of steel sheet metal homemade arrowheads set into slotted notches in the shafts and wrapped with thread, 4) flint arrowhead attached the same as with #3, but wrapped with dental floss, and 5) bone arrowhead attached with imitation sinew.

putting points on our arrows. For target practice or hunting small animals, a simple tapered point on the shaft might suffice. But for a more lethal type of arrowhead, we might attach a triangular-shaped point cut from sheet metal and filed sharp; a stone point

knapped from flint, obsidian, or glass (for instructions on flint knapping, see *Makeshift Workshop Skills for Survival and Self-Reliance*); a pointed tip shaped out of a hard piece of bone or even hard plastic; or if we're especially lucky we might even have some commercially manufactured broadheads available to us. Most homemade arrow points aren't legal for hunting in certain jurisdictions, but that wouldn't be a consideration after an apocalypse.

We should also remember that the more uniform we make the size and weight of our arrows (including the shaft length and diameter, weight of the points, and fletching, etc.), the more consistent will be our accuracy with them.

An Experimental Quickie Bow, Made Quickly

For a recent experiment to test how fast I could construct a functional quickie bow, I cut a 5½-foot-long branch of about 1½-inch diameter at its thickest end from the lilac tree in my front yard for the bow, and another fairly straight 3-foot-long skinny branch from the same tree for the arrow, and timed my progress.

I used a length of parachute cord for my expedient bowstring, attaching it to notches I had cut into both ends of the stick (bow stave) using a pruning saw. I used a hangman's knot to attach the cord to the stave at the permanent end and a timber hitch at the other end. I like using the timber hitch for this because it is quick and easy to form, holds securely under tension, and is adjustable for variable bowstring length as desired. I tapered the front end of the arrow shaft into

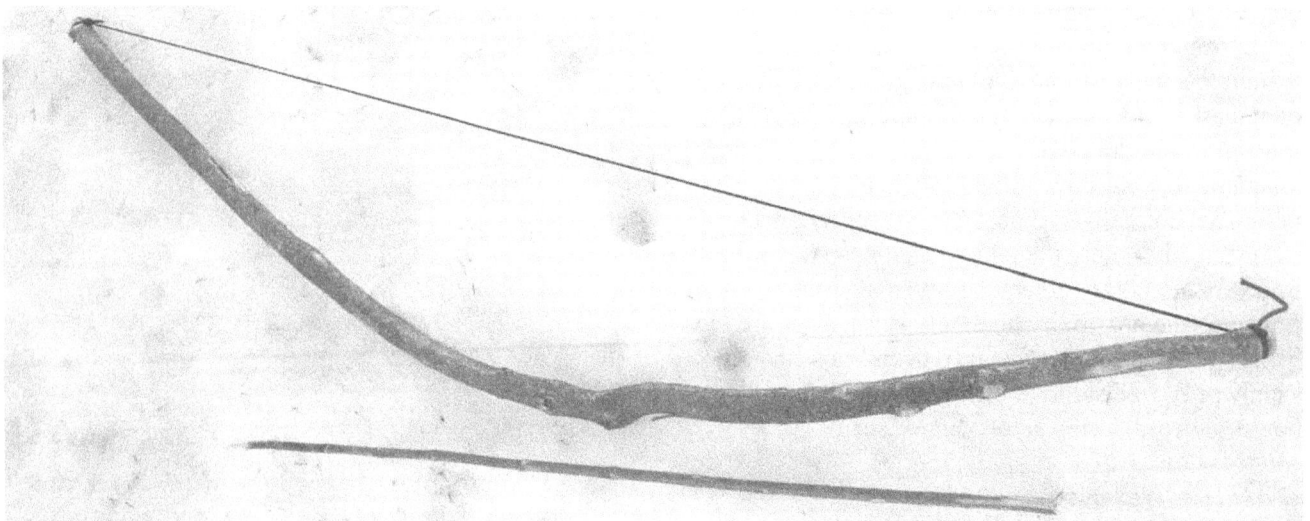

Quickie bow and quickie arrow, fabricated quickly.

The author drawing the quickie bow with the un-fletched quickie arrow.

a crude point and made a notch in the back end with the pruning saw to form an expedient nock that fit the thick bowstring.

For this experiment I did not bother to attach any fletching to the arrow, as I was trying to simulate the most rudimentary, quickest-to-construct, but still potentially lethal close-range weapon that someone might be able to build in a pinch, with only a piece of cord and a cutting tool of one kind or another. The draw weight was not precisely measured in this crude experiment, but my guess would be something in the neighborhood of 30 to 40 pounds. I was able to sink that unfletched arrow deep into a hard-packed snow-bank from about 29 feet away.

The length of time from sawing the branches off the tree until the photo of me aiming the arrow was taken (by my son, Eugene, who is not yet seven years old as I write this) was 47 minutes.

Naturally, more time spent refining this equipment could have yielded greatly improved results, with more and possibly straighter arrows having fins and sharper arrowheads, a bow in proper tiller strung with a more efficient string, and so on. But this experiment does reveal just how quickly a dangerous weapon could be devised with only a minimum of tools and materials.

The Quickie Bow Starter Kit

It is true that wherever trees and shrubs grow, usable bow staves and arrow shaft materials can be found and harvested in a survival situation, and usually fairly quickly. But because bird feathers aren't particularly easy to find in sufficient quantity to fletch a batch of arrows in a hurry, and because constructing a serviceable bowstring from natural materials without first killing a deer and harvesting its tendons can present a dilemma, I decided to assemble my own lightweight, compact, quickie bow starter kit to go in my survival pack.

The key to this kind of a kit is to include mainly the archery components that are the hardest to find or fabricate quickly in nature. For example, knapping arrow points from flint or pieces of glass is not necessarily difficult for someone who has mastered the skill, but finding the required materials and shaping the points in a timely manner may not always be achievable. Therefore, I include several manufactured steel broadheads in my kit that will quickly twist onto the ends of expedient arrow shafts in an emergency.

I included half a dozen factory-made bowstrings in this kit. My rule of thumb is to select somewhat heavier strings than I might need (I selected strings for 50 to 70 pounds of draw weight), that are also a bit longer than probably necessary. My reasoning is that a longer, heavier bowstring, while maybe not always ideal, will accommodate more possibilities of bow selection. I may hope that my future expedient bows will have a draw weight somewhere between 40 and 60 pounds, but who knows what kinds of wood or staves I might find in nature to work with when the time comes? A string that is too long to fit the bow I build can usually be shortened by twisting it several turns, but I don't know any practical way to make a string that is too short any longer. Likewise, a string that is too heavy will still bend a weaker bow, but a lightweight string on a heavier bow might break. So I err on the side of extra, given all of the unknown variables I will have to face later.

I did heed the wisdom of tying little tags to the bowstrings that note the length and rated draw weight of each, to eliminate having to guess later on. My kit also includes a small beeswax candle for keeping the strings waxed. All archers know that waxed strings give superior performance and last a lot longer than unwaxed strings.

Arm guards and finger shooting gloves don't add very much weight to the kit, and both items can make archery shooting a lot easier, so I consider them worth their space. Preshaped feather vanes that are ready to attach to arrows (they can be either tied or glued onto the arrow shafts) are ridiculously lightweight but will

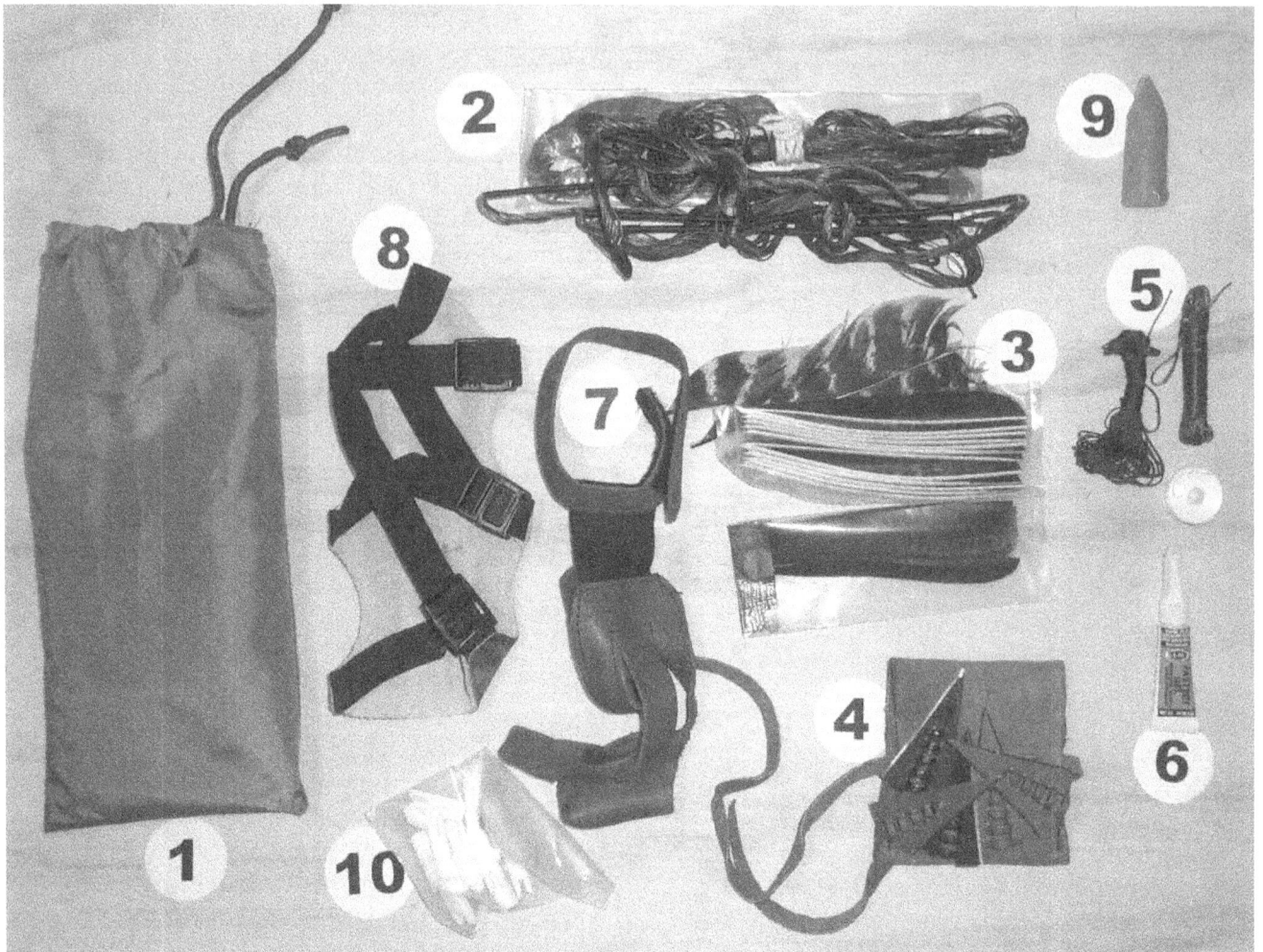

This quickie bow starter kit includes: 1) nylon stuff sack to house components, 2) several factory-made bowstrings of various lengths and draw weights, 3) pretrimmed feather vanes, 4) steel broadheads with their own leather pouch to protect their points, 5) miscellaneous thread and extra B-50 Dacron for repairs and homemade strings, 6) small tube of archery glue, 7) shooting glove, 8) arm guard, 9) beeswax for string maintenance, 10) plastic arrow nocks.

enhance the equipment dramatically, so I included enough for several arrows. I put in a tiny bottle of archery glue for gluing on feathers and nocks to save time and frustration later on, and I also included small-diameter cord (including B-50) and strong thread for tying on fletching and arrowheads, or for bowstring repairs.

My entire kit weighs only about 10 ounces and doesn't take up too much space in the apocalypse-ready kit.

Crossbows

Our fourth and final category under the general topic of archery is the crossbow. As most readers well know, a crossbow is a projectile launcher best described as a bow mounted to the front of (usually) a

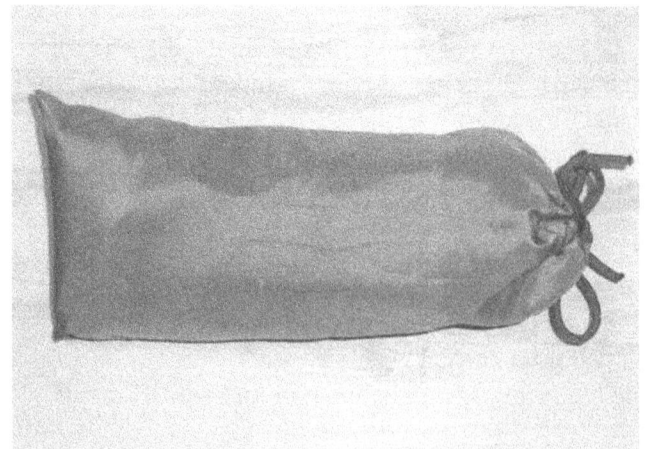

The entire quickie bow kit stores inside this 4 x 9-inch nylon stuff sack and weighs just 10 ounces.

shoulder stock, which is aimed and shot like a firearm.

Crossbows—the medieval crossbows commonly referred to as *arbalests*—have been around for centuries and, for the specialty purposes they fulfill, merit our attention. Unlike other types of bows that require human muscle to draw and aim, crossbows don't have to be made ready just before taking a shot or held ready with great effort at full draw until the opportunity for the best shot presents itself. Instead, they can conveniently be kept cocked and ready to shoot almost indefinitely until the important shot needs to be taken. In this sense, they simulate a firearm more than a bow.

Crossbows are somewhat common today and come in a variety of configurations. Just as with regular recurve and compound bows, a bargain hunter may find usable crossbows priced reasonably at yard sales or through online outlets such as Craigslist. These will vary a great deal in usability and should be inspected carefully before purchase.

There are some characteristics shared by most variations of crossbows that I think the readers ought to be aware of. To start with, the arrows shot from crossbows are more often referred to as "bolts" (commonly called *quarrels* in medieval England). Bolts are typically much shorter than arrows, and unlike conventional arrows, most lack any significant nock at the back of the shaft.

The limbs on crossbows (commonly called "prods") are typically much shorter than those on conventional bows. A crossbow that measures three feet across its limbs (as does the Exomag from Excalibur Crossbows) is considered a very large

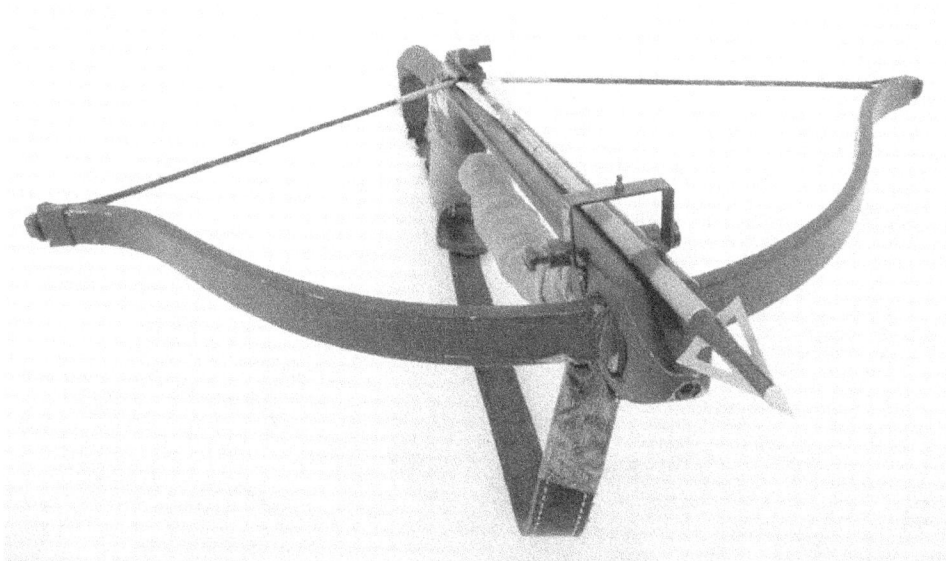

The lethal nature of a crossbow is obvious in this photo.

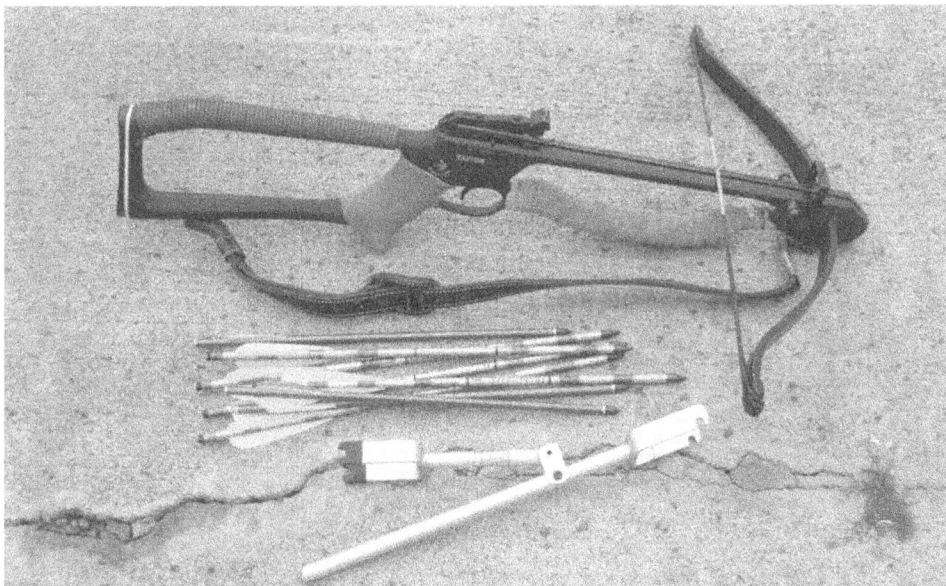

The Barnett Panzer—shown here with its "goat's foot" cocking lever and bolts with target points—is a hunting recurve crossbow having 150-pound draw weight. I wrapped the aluminum fore-end with a leather strip for a better, more comfortable grip in winter.

Two older-style crossbows of the variety typically found at yard sales. The bow at bottom is possibly home-made and lacks an arrow hold-down.

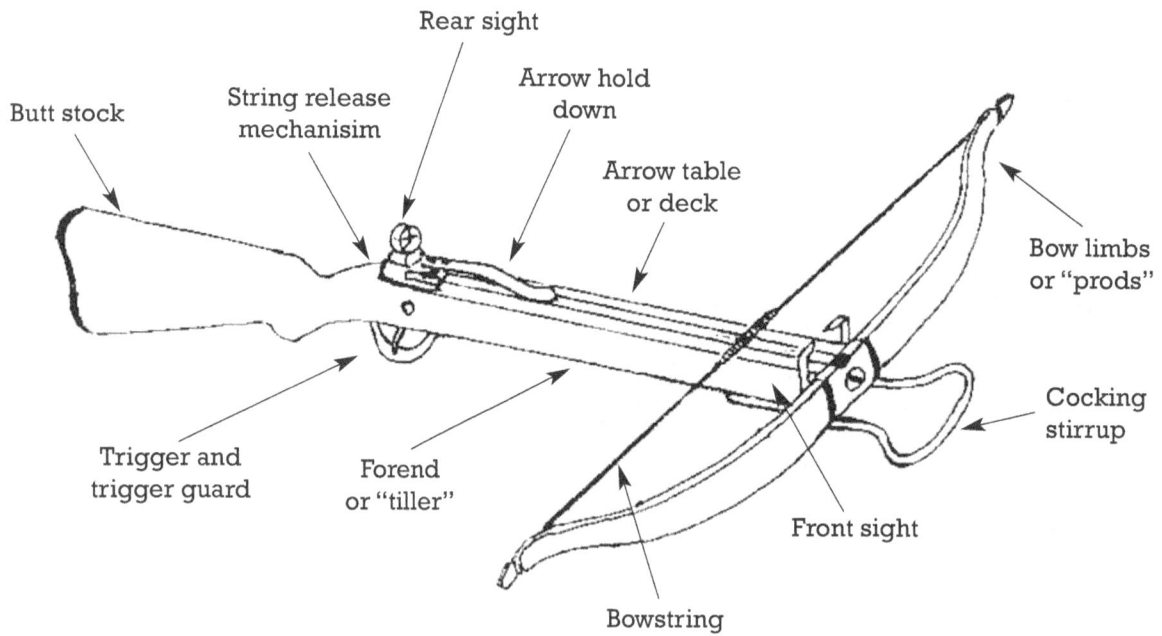

Common Crossbow Terms

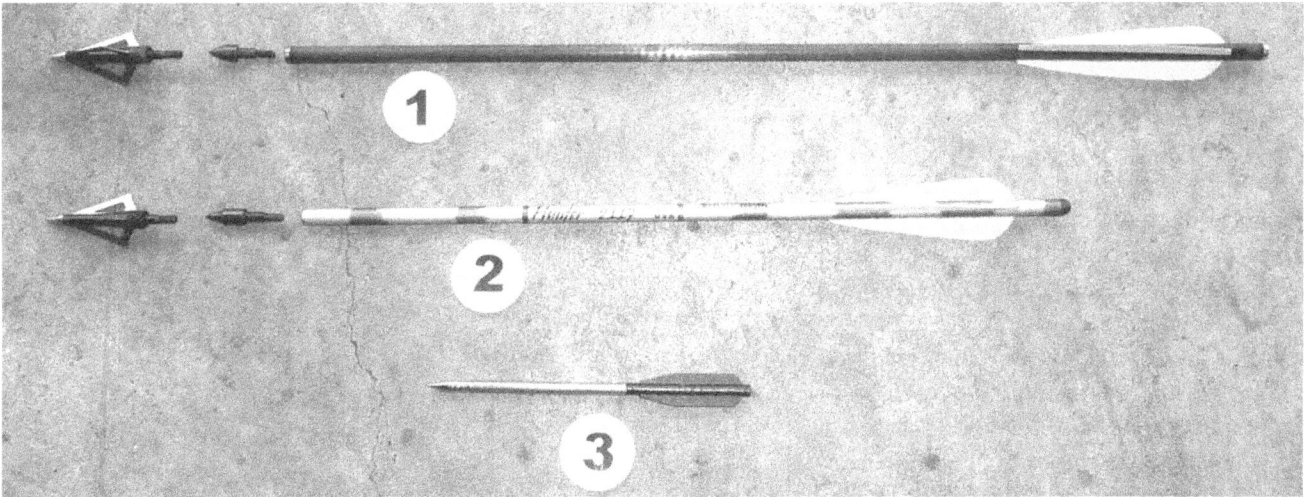

Three different projectiles for crossbows: 1) 20-inch aluminum arrow intended for the wide-limbed Excalibur brand crossbows, with two optional point types, 2) 16-inch bolt with two optional point types for most contemporary hunting crossbows, and 3) 6-inch dart/bolt for pistol-type and very small crossbows.

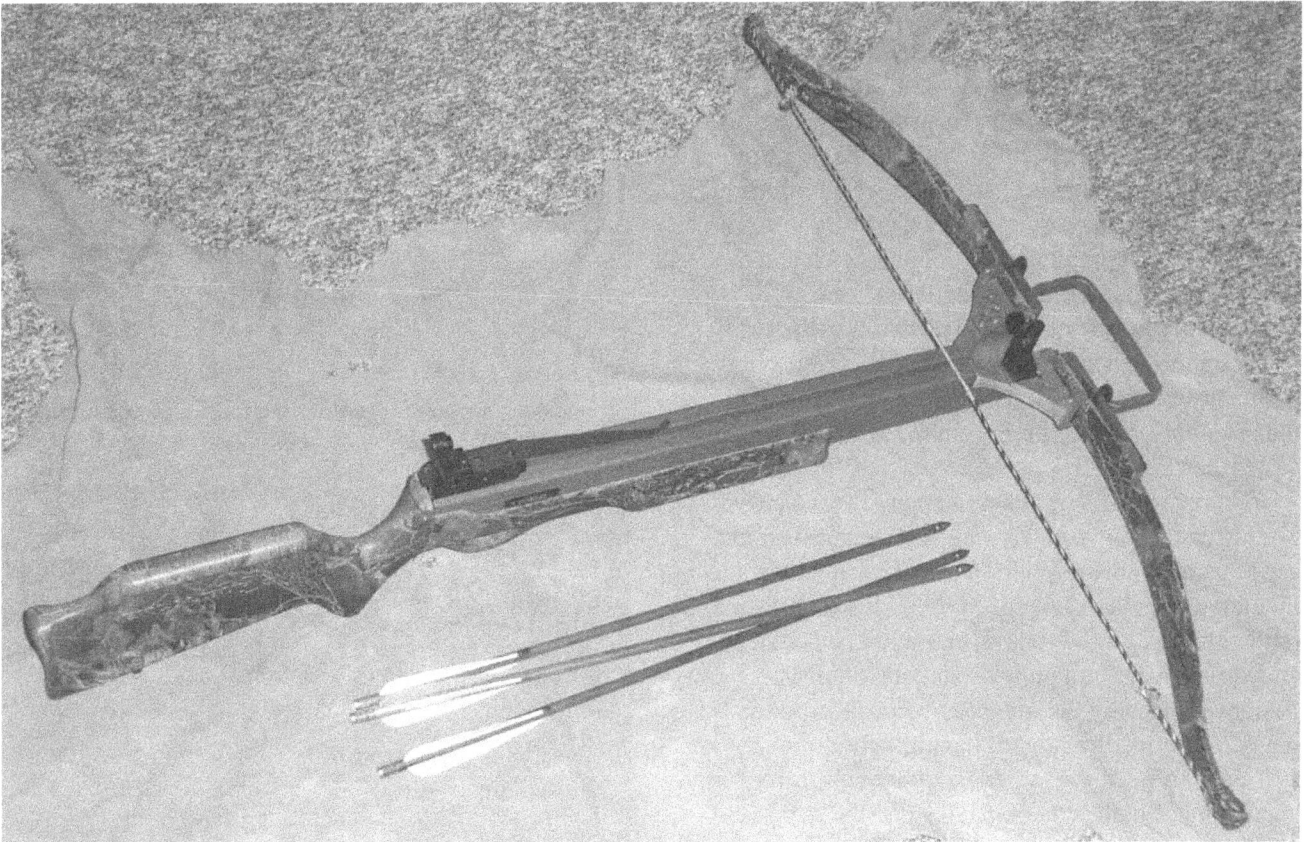

Top of the line in modern hunting crossbows is Excalibur's Exomag, a recurve crossbow that draws 200 pounds and launches 20-inch arrows at over 300 feet per second.

crossbow. By contrast, most recurve and longbows for adult archers will be longer than four or even five feet, nock to nock.

Bows mounted on crossbows are configured as flat bows, longbows, and recurve bows, and nowadays compound crossbows are common. One distinct advantage of the compound limb is that the trigger mechanism doesn't hold the full draw weight of the bow and is therefore subject to less stress. The trade-offs are that string changes are much more difficult with compound crossbows than with other types, and there is some added weight with those typically larger limbs, pulley cams, and cables.

Draw weights are typically much heavier on crossbows than on conventional bows, ranging anywhere from 50 pounds on the lighter end to 175 pounds, and a few models (like the Equinox from Excalibur) will draw as much as 225 pounds. The Exomag from Excalibur that I bought eight years ago draws 200 pounds.

For this reason, and also because the string brushes over the track during use, crossbow strings sustain enormous shock and wear and have a shorter life than those on conventional bows, even though they are thicker and heavier. Bill Troubridge at Excalibur Crossbow explained that shooters should expect 300+ shots with their bows before having to replace the string. I should stress that it is essential to keep strings well waxed and avoid dry firing bows (i.e., shooting without a projectile), which can break a string as well as cause damage to the bow.

Probably the easiest way to string a conventional flat bow or recurve crossbow, without using a stringer aid of one type or another, is to brace one limb firmly against the floor with one loop of the string secured on it and, while keeping the bow pointed down and the limb braced against the floor under the bottom of your boot, simply lift the other limb up and slip the string's free loop into position on the free nock. If your bow has a stirrup, you can step your foot into that rather than stepping directly on the braced limb. It's actually easier than it sounds.

There are different ways to span (cock) a crossbow. Medieval crossbows were spanned using a variety of gadgets that provided a mechanical advantage, including windlasses, cranequins, goat's foot levers (still commonly used today), belt-and-claw devices, and various cord and pulley spanners. Some crossbow designs, especially the inexpensive pistol types and many of the smaller models having light draw weights, and even a few of the larger exotic models, incorporate some type of built-in cocking device.

Probably the simplest method for spanning a crossbow (but only by shooters without major back problems, and with bows that possess a stirrup) is by pulling up on the string with both hands. This is accomplished by stepping into the stirrup, grasping the string on both sides of the track with the bow pointed downward, and lifting the string to cock the weapon. I find this method fairly quick and easy to

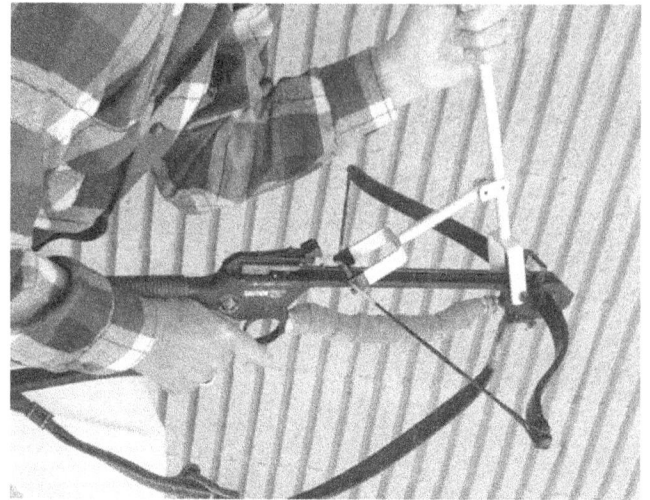

Spanning the Barnett Panzer using a modern goat's foot lever.

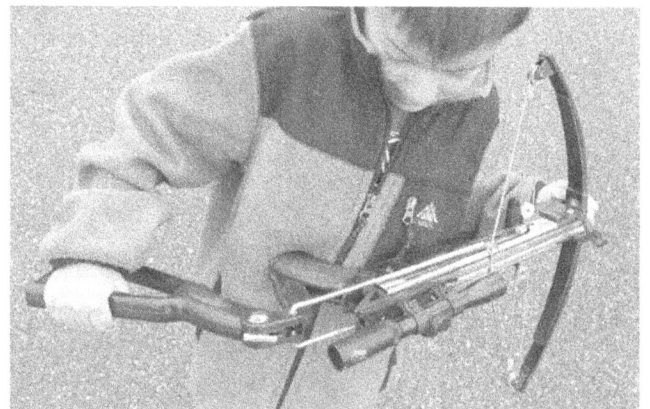

Spanning a small crossbow using its convenient built-in cocking mechanism.

Spanning a recurve crossbow having 200-lb. draw weight simply by lifting up on the string. Leather gloves can make this task easier on the fingers.

The Excalibur hunting crossbow features an aperture rear sight and a fiber-optic front pin.

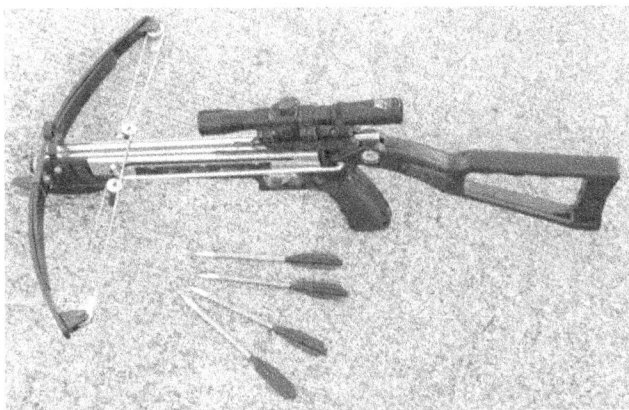
The little Chinese-made FX-II is an inexpensive crossbow that wears a telescopic sight.

Aiming the scoped crossbow.

accomplish, especially when wearing leather gloves, even with the 200-pound draw weight of my Excalibur recurve crossbow.

Whenever using this method, take care not to let the stirrup slip out from under your shoe or boot while spanning the bow. Also, it is important to apply an equal amount of pull to the string on both sides of the track in order to maintain proper alignment for the most accurate shot.

Most modern crossbows are equipped with sights of one form or another, typically resembling those found on firearms. Some models can be fitted with scopes, although the crossbow is not really a long-range weapon. Sixty yards is about the outer limit of crossbow effective range, even with the best of them. Even so, surprising accuracy is achievable within this range.

Safety with a crossbow is paramount, as these weapons function by releasing an enormous amount of stored energy. I learned the hard way about the hazard when I attempted to launch a lightweight 6-inch bolt intended only for the small pistol-type crossbows from the track of a larger, more powerful model. The lightweight projectile doesn't provide the appropriate load (resistance) for the power of the larger bow, and the desired stability is lost. In my case, the arrow flipped off the track and flew in an unpredicted direction, while the bow sustained an excessive amount of shock, as if it were dry fired with no projectile at all. A great way to damage a crossbow and endanger yourself or any bystanders at the same time! It is very important to use a bolt size and weight appropriate for the size and power of the bow.

The only thing that holds the bolt in position on the track of a ready-to-shoot crossbow is a simple little device called an arrow hold-down arm, which is really nothing more than a curved flat spring shaped like a belt clip that presses down on the back of the bolt while it's on the track. I have seen homemade crossbows that lacked this essential component, but I can't imagine how a crossbow would be safe to use without it.

Another important safety tip: keep the fingers of the hand supporting the forearm while aiming below the track line and safely clear of the bowstring during launching.

The powerful hunting crossbows generate a noticeable back kick upon launching a bolt; you can feel the powerful thrust of the process, but in my experience the recoil is mild compared to that of just about any big-game hunting rifle. Rather than a hammering jolt against the shoulder, the recoil is more of a push.

For information about the latest quality products in this category, I recommend Barnett Crossbows (www.barnettcrossbows.com) and Excalibur Crossbows (www.excaliburcrossbow.com).

A fascinating, heavily illustrated book for readers interested in learning more about ancient and medieval crossbows is *The Book of the Crossbow* by Sir Ralph Payne-Gallawey, originally published by Longmans, Green, & Co., London, in 1903 and followed by a Dover Publications reprint edition in 1995.

Crossbows can be very effective weapons, but they do have a few drawbacks when compared with most of the other weapons we have considered for the apocalypse. These include their awkward and bulky shape, limited effective range (again, up to about 60 yards realistically), and slow rate of fire. Even a conventional bow can usually be drawn, aimed, and accurately shot more times in a given span of time than can a crossbow.

CHAPTER 10

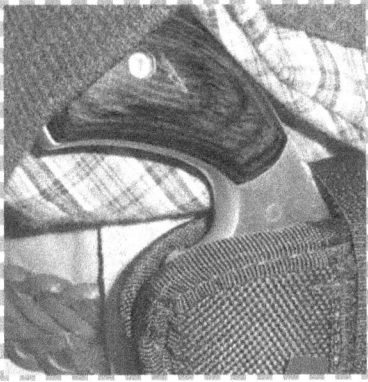

Selecting Contingency or Backup Firearms for the Apocalypse

In the previous chapter, we considered alternative types of weaponry, or "non-firearms," as secondary weapons to our conventional firearms. In this chapter we will look at different conventional firearms that we might wish to maintain as backups to our primary firearms for the apocalypse. It is worth considering that a backup firearm could easily be a lifesaver for us in a desperate, postapocalypse emergency situation.

The guns we will talk about here will conform to a different set of priorities than we demand of our primary doomsday firearms. For one thing, they can't weigh very much, because just our primary weapons alone will tend to weigh us down more than we will probably desire. Any backup firearms we carry will have to be small and lightweight enough that they won't interfere with our mobility or our ability to carry enough of the more essential primary gear. So this doesn't allow us much in the way of performance-oriented features. Their intended purpose is restricted to close-up self-defense.

The over-under "double derringer" that became popular in the late 1800s was a conveniently compact pocket pistol that enjoyed a degree of popularity, famously with riverboat gamblers, and it remained in production from 1866 until 1935. Its .41-caliber rimfire cartridge was notoriously underpowered for a defense round, however. According to Wikipedia, this cartridge was loaded with a 130-grain lead bullet over 13 grains of black powder that attained a muzzle velocity of 425 fps and generated a mere 52 ft-lbs. of energy—less energy, in fact, than either a .22 Short or

Two early hammerless pocket automatics for emergency backup: Colt's .380 on the left, FN Browning .32 ACP on the right.

The original two-barreled Remington Derringer from the riverboat gambling days is a very traditional type of "belly gun." Its .41-caliber rimfire cartridge is not as powerful as even the most marginal pocket gun cartridges of today.

.25 Auto bullet! Original derringers now carry a fairly high collector value, and a reliable source of the loaded .41 rimfire ammunition could be difficult to find these days, so this is probably not a very practical weapon for anyone seriously preparing for the apocalypse.

Even so, the basic design of the double derringer is alive and well today in various forms and certainly viable as a backup weapon with the right modern ammunition. Different versions of this basic style of pistol have been manufactured in recent decades by several arms makers and in a wide variety of chamber offerings.

Originally characterized by their short over/under barrels, bird's head grip shape, unguarded spur triggers, single-action mechanism, and conveniently small size, the derringers that have appeared in recent years are usually heavier, much stronger, chambered for a variety of the more powerful common handgun cartridges of today, and available in several barrel lengths and trigger configurations. The Texas Defender from Bond Arms (bondarms.com), for example, comes with 3-inch interchangeable barrels, is available with 13 caliber options from .22 LR to .410 shotgun (including .357 Magnum and even 10mm Auto), is constructed of stainless steel, has a trigger guard, and weighs 20 ounces.

The American Derringer Corporation (www.amderringer.com) similarly offers an interesting variety of two-barreled derringer-style pistols, including their Lightweight and Ultra Lightweight models that weigh only 7.5 ounces. These guns are styled very close to the original Remington over/under derringers and are available in seven chamber options, from .22 LR to .44 Special. American Derringer also offers its DA38 Double Action, which is a two-shot pistol available in either .38 Special or .40 Smith & Wesson, weighs just 14.5 ounces, and has a double-action, partially guarded trigger. They even have a lighter version of this pistol in .22 Magnum.

An incredibly compact two-barreled over/under derringer-type pistol chambered for .45 ACP is the DoubleTap from Heizer Defense (www.heizerdefense.com). The pistol has a titanium frame with rounded-over edges so it won't snag on clothing, weighs only 14 ounces empty (also available with an aluminum frame at 12 ounces empty), is just 5/8-inch wide, and is advertised as "the world's smallest and lightest .45 ACP concealed carry pistol on the market today." The pistol is also available in 9mm.

The three features I like best about this little gun, besides its convenient size and the power of its cartridge for defense purposes, are that it comes with optional

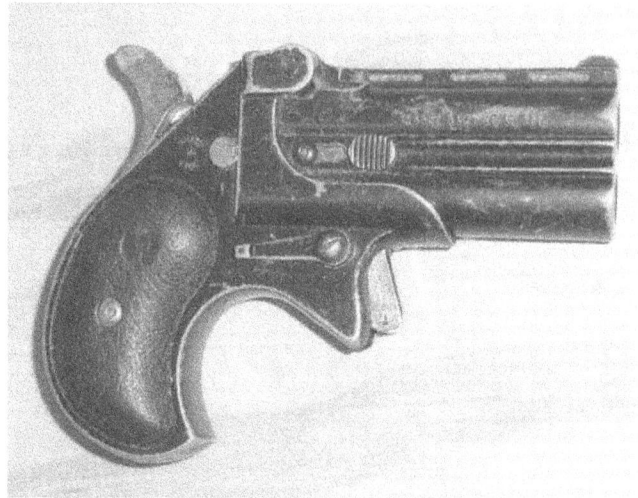

An over-under derringer in .38 Special from Davis Industries, which is no longer in business.

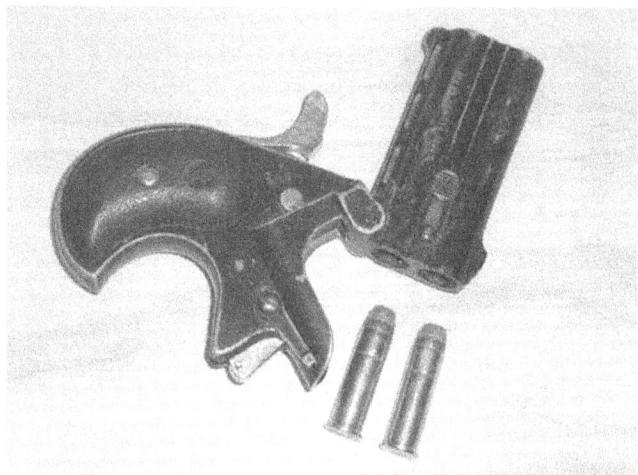

Showing how the barrel of the Davis derringer tips up for loading and unloading.

The five-shot, single-action .22 Mag Mini Revolver from North American Arms.

The DoubleTap .45 ACP derringer from Heizer Defense. Image courtesy Heizer Defense, LLC.

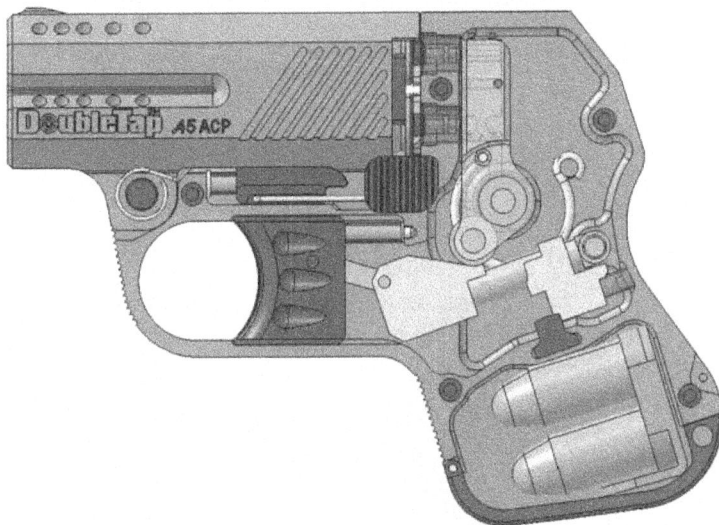

The internal workings of the DoubleTap pistol. Image courtesy Heizer Defense, LLC.

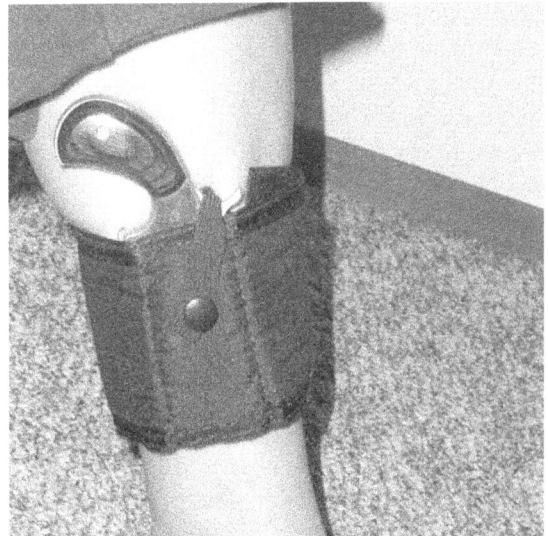

The North American Mini hides comfortably in an ankle holster.

ported barrels to reduce recoil, it has a unique patent-pending ball bearing double-action trigger system, and it will store two spare loaded cartridges in its grip.

With the DoubleTap pistol, the shooter need not worry about having to cock a hammer before pulling the trigger because this gun is double-action. And if two quick .45 rounds can't tilt the outcome of the fight in your favor, then you are simply in over your head!

The smallest revolvers in the world, as far as I know, are the famous Mini Revolvers from North American Arms (www.northamerican-arms.com), the makers of the popular Guardian .32 ACP pocket auto pistol. A number of Mini Revolvers are featured on their website, with different barrel lengths, grip configurations, and sighting arrangements, but a very common (and very tiny) model is the five-shot, stainless steel, single-action revolver chambered for .22 Magnum and having a 1 1/8-inch barrel and spur trigger. This gun weighs only 5.9 ounces unloaded. A gun this small is very easy to carry comfortably and conceal. I even showed how it could be stored in a hidden cavity inside a hardcover book in *More Makeshift Workshop Skills*.

The cylinder of the North American Mini is emptied or unloaded by using its cylinder pin, which must be removed from the gun anyway for loading. One of my favorite features of this gun's design (besides its incredibly small size) is the slots in the back of the cylinder between chambers that allow the hammer to rest down safely between cartridges while the gun is carried. This way a hard blow to the back of the hammer cannot fire the gun accidentally, even with the cylinder full of loaded rounds.

I was pleasantly surprised when I learned that the Mini revolver could be fired with some degree of accuracy beyond 10 feet. A full-sized galvanized steel trash barrel I found at a local shooting area that had already been shot full of holes was the perfect target for the tiny gun at 35 yards, because I was able to hear when the bullets hit the metal side. This gun has no rear sight to speak of, but I discovered that by holding the top of the barrel level with the entire front blade showing over the frame and aligned with the trash barrel's center of mass, it was easy to hit such a sizeable target at that distance with every shot.

I do realize that a .22 Magnum bullet fired from an inch-long barrel is not going to deliver an enormous amount of kinetic energy at the target, and especially not at 35 yards, but the important thing to reiterate here is that these tiny guns are strictly last-ditch, close-range weapons. We put them into action whenever our first-choice weapons are no longer serviceable or available to us when we need them.

As indicated, the Mini revolvers are available in different versions. One model with a folding handle allows the gun to be conveniently stored and well protected in the closed arrangement. When the handle is opened, it forms a larger, more comfortable shooting grip.

The snub-nosed revolver—typically a double-action revolver with a small frame, minimal fixed sights, and barrel length of 3 inches or less (2 inches is the most common)—has been a popular backup firearm for off-duty police and undercover federal agents for decades. The most traditional caliber for snub-nosed revolvers, in America anyway, is the .38 Special, but today "snubbies" can be found chambered for such other rounds as .32 H&R Magnum and .357 Magnum.

Exposed hammer spurs are sometimes cut down on small double-action revolvers to make them less likely to snag on clothing when pulled from pocket liners and other small spaces. The spur on the hammer is not used with a gun in double-action mode, and for that reason you can even find hammerless versions of double-action-only snubbies having no exposed hammer at all.

Because these compact revolvers are so much lighter than full-sized guns, their recoil with any given cartridge is going to be greater than that of any larger gun in the same caliber. High-energy magnum rounds are therefore not nearly as common in the snubbies as are more moderately powered revolver cartridges like the .38 Special.

The popular trend with snubbies nevertheless seems to be toward smaller handguns composed of lighter materials like aluminum, titanium, and even

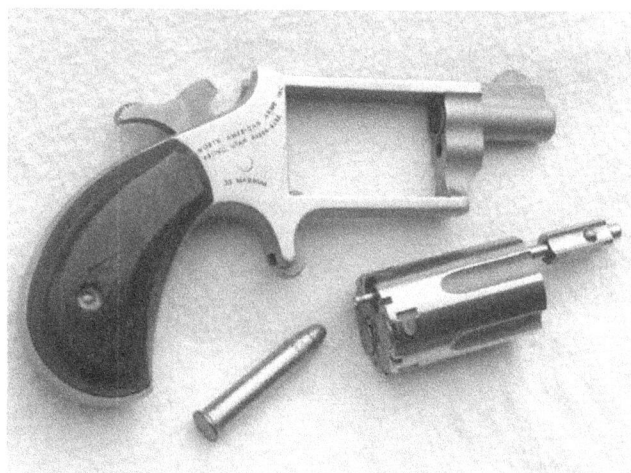

The cylinder pin of the North American Mini is used to unload the cylinder, after being removed from the gun.

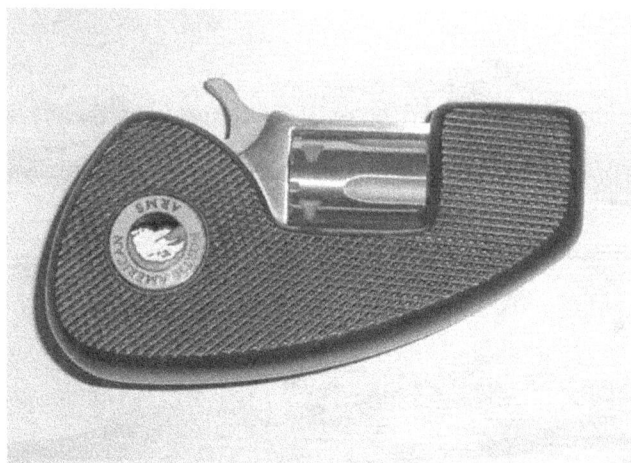

North American Mini revolver with its grip handle folded up for storage.

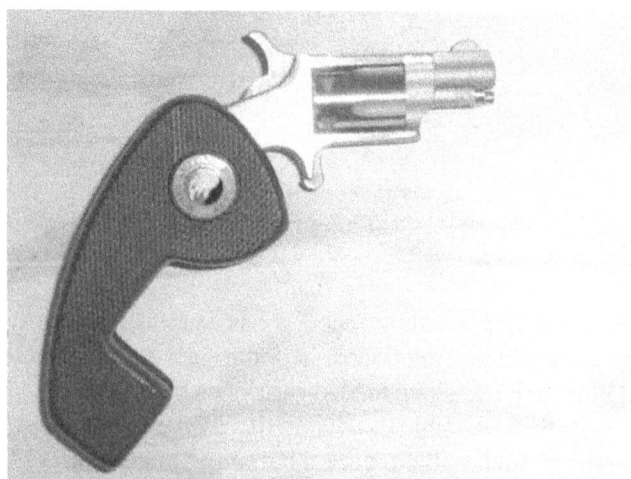

North American Mini with its handle extended for shooting. This particular gun is chambered for .22 LR.

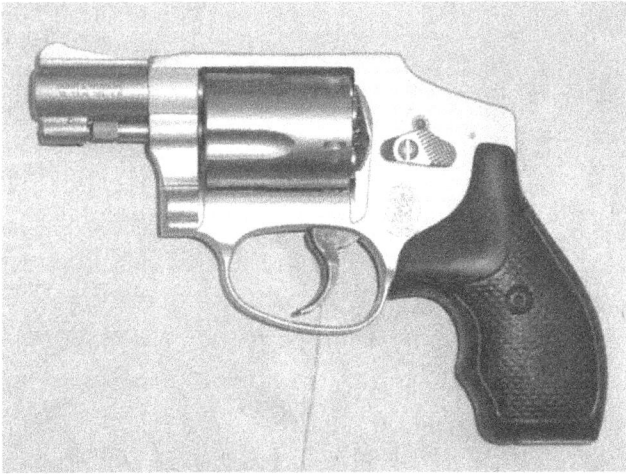

A very lightweight, snub-nosed .38 Special hammerless revolver from Smith & Wesson.

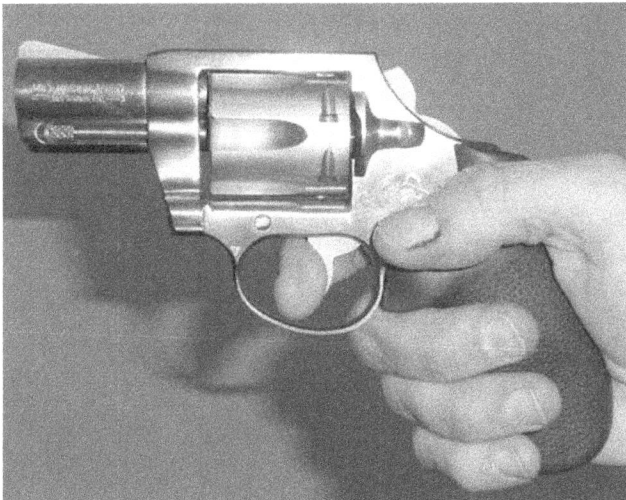

Colt's Magnum Carry is a six-shot, double-action, stainless steel snub-nose in .357 Magnum. This one has a customized "bobbed" hammer spur, making it more suitable for carrying in a pocket.

Ruger's LCR double-action-only snub-nosed .38 Special.

Ruger's LCP .380 ACP automatic in the palm of a lady's hand.

polymer, as is the material used for the grip and trigger housing of the Ruger LCR (Lightweight Compact Revolver), making this particular gun at 13.5 ounces possibly the lightest snub-nosed .38 Special revolver on the market. This is a result of the growing popularity of concealed carry in America, where more and more people are concerned about their personal security than ever before and want a gun that is easy and unobtrusive to carry.

Within this same niche, we find the smallest of the pocket automatics available today, in calibers from .25 Auto to .380 ACP. Colt's original Vest Pocket Model of 1908 in .25 ACP was a popular tiny gun in its day, but its .25-caliber bullet has never been considered a reliable defense round by almost any standard. Its most common 45-grain bullet emerges from the little pocket pistol's barrel with only 66 ft-lbs. of kinetic energy—less than even the little .22 Short's usual 77 ft-lbs.

However, the next size up is the .32 ACP (known in Europe as 7.65mm), and modern loadings of this slightly heavier round generate around 125 ft-lbs. from a pocket pistol, making this one a tad more respectable. There are a number of small auto pistols on the market today chambered for the .32 ACP.

The smallest of the pocket automatics made by Beretta are quality made, very small, and incorporate the unique tip-up barrel feature in their design that allows the shooter to quickly load or unload a round directly into or from the chamber without having to work the gun's slide. These models include the Bobcat chambered for either .25 ACP or .22 LR and the Tomcat in .32 ACP.

Among the most famous of the small auto pistols are the Walther PP, PPK, and PPK/S pistols. This is largely due to the PPK being the weapon of choice for Agent 007 in the James Bond film series.

These little Walther pistols have been chambered in .32 ACP and .380 Auto (known as 9mm Kurz in Europe). The 20-ounce PPK is the smaller version of the PP that appeared two years before it in 1929, one of the first ever double-action auto pistols. The PPK/S is of slightly larger dimensions than the PPK to conform to the 1968 U.S. importation requirements, and it weighs an ounce and a half more than the PPK. The magazine capacity of the PPK/S was also increased by one round more than that of the PPK. Both pistols are noticeably smaller than the PP model.

The original Walther PP double-action design was clearly the inspiration for several pistols that followed it, including the Czech CZ 50 chambered for 7.65mm and the Russian Makarov pistol chambered for 9x18.

As of this writing, both the PPK and PPK/S models are available from Smith & Wesson in the United States, and they can be viewed on the company website (www.smith-wesson.com).

Recently, the most concealable 9mm pistols seem to be getting smaller as well. Ruger's LC9, for example, is a compact 17.1-ounce 9mm auto pistol with an overall length of 6 inches, glass-filled nylon grip frame, and magazine capacity of 7+1 rounds. This gun is almost three ounces lighter than the popular subcompact GLOCK G26 9mm without its magazine.

Kimber recently introduced its Solo 9mm, which is already becoming incredibly popular within this compact 9mm automatic pistol niche. It was named "Handgun of the Year" for 2012 by *American Rifleman* magazine for its ergonomics and performance. This "micro-compact" is a 6+1 capacity, single-action, striker-fired automatic that weighs only 17 ounces and features a lot of no-snag rounded edges.

The new CM9 from Kahr Arms is another very compact 9mm semiauto with a polymer frame, 3-inch barrel, and 6-round magazine. It weighs just 14 ounces empty, and its width is only 1.12 inch. Probably the best thing about this new double-action-only 9mm, besides its compact size and general high rating, is its retail price—under $600 in 2012.

An amazingly lightweight and compact .380 is Ruger's LCP, which is a 9.40-ounce pistol with a glass-filled nylon grip frame and cartridge capacity of 6+1. A gun this small will hide easily inside almost any pocket, purse, fanny pack, or glove compartment until an emergency arises and it is suddenly needed.

There are a number of holster options for the tiny

Beretta's very compact Tomcat pocket auto, chambered for .32 ACP.

Showing the tip-up feature of Beretta's .32 Auto pistol.

Walther PPK/S pistol in caliber 9mm Kurz, which is the same as .380 Auto.

A 9mm round (left) next to a .380 ACP for size comparison.

The Ruger LCP in a pocket holster.

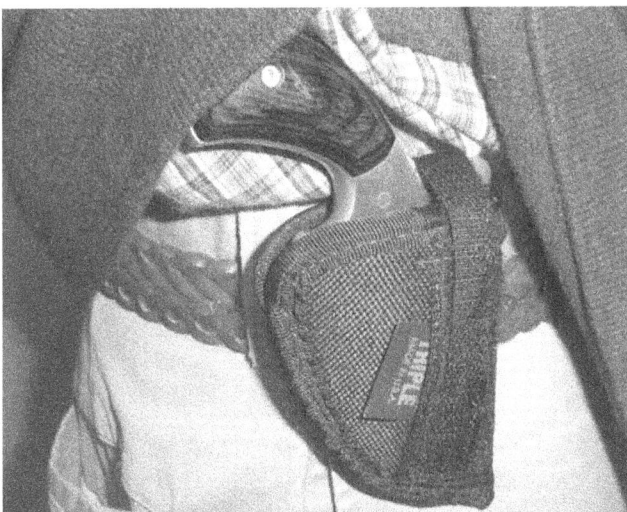

The tiny Mini Revolver is easy to conceal under a sweater while riding in a simple hip holster.

guns. The smallest guns will ride comfortably in an ankle holster, in a bellyband (hence the term "belly gun"), in a shoulder harness, inside the pants, inside a pocket holster, or in a simple traditional belt hip holster. Fortunately for those on a tight budget, it doesn't take an expensive carry rig to do the job.

Guns that are going to be carried inside a pants pocket should be inspected and cleaned often, because they tend to collect a lot of lint and dirt from the cloth liner and all the other things routinely carried in a pocket. For this purpose, the inside-pocket type of holster makes a lot of sense, because it keeps the gun supported above most of the loose coins, keys, handkerchiefs, and lint that tends to accumulate in the bottom of the pocket.

Sometimes while packing the smallest handguns, it is very easy to forget you are armed, because the tiny weapons tend to stay comfortably out of sight and out of mind. You don't want to attempt to enter a restricted area forgetting that you're armed, or unknowingly leave a loaded gun in a coat pocket where it could be discovered by a child. Make it a habit, even a ritual, to consciously secure your carry gun at the end of the day.

APPENDIX

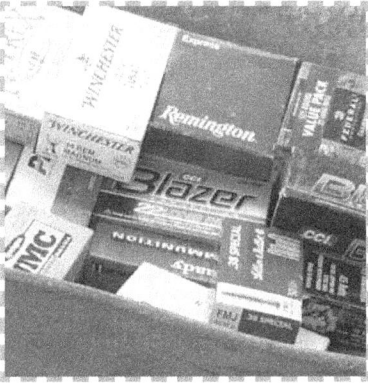

A Few More Useful Arming Tips and Ideas

For riflescopes, the clear, see-through variety of cover allows a scope to be used while providing some protection to the optic at the same time.For semiautomatic weapons supported by multiple magazines, marking the individual magazines makes it easier to distinguish one from another. Then if one develops any problems at some point, or if you rotate loaded with unloaded magazines, marked (ideally numbered) magazines will be easy to identify and monitor without confusing one with the others.

It can be handy to keep a small waterproof vial containing cotton balls with firearms. The cotton can be stuffed in your ears as poor man's earplugs on the day you failed to bring your range bag containing your regular shooting muffs, or it can be used as fire-starting tinder in the woods during a survival emergency.

The correct sizes of hex or Allen wrenches (and possibly small screwdrivers as well) that fit the screws on your apocalypse weapons can be adapted for carrying on a key chain. You can usually heat one end of the tool with a torch until it's soft and then bend it into an eye loop using needle-nosed pliers such that it can then ride on a key chain.

New guns, especially semiautomatics with their characteristic level of friction in moving parts, can occasionally be too stiff to function smoothly at first. It is not uncommon to experience some malfunctions with a new gun until the parts work together for a while. Often, the best cure is simply to take the gun shooting and break it in. A few shooting and cleaning sessions can do wonders for a new gun. Besides, who

Two GLOCK magazines engraved with numbers to help distinguish one from the other. The soft plastic was easily engraved with a narrow gouge chisel.

wants to face the apocalypse with an untested or unfamiliar weapon?

Always remember that the direction you move a rear sight, either to one side or up or down, will be the same direction that the bullet's impact will move. If you wish to move your shot groups to the left, then simply adjust the rear sight to the left, and so on.

A shallow metal cookie sheet or baking tray makes an excellent receptacle for the tiny screws, pins, sights, springs, and other separated small parts whenever completely disassembling a firearm for repairs or extensive cleaning. The shallow container allows quick visibility of and convenient open-top access to the tiny parts,

A wine bottle cork keeps dust and moisture out of this musket's bore.

while at the same time preventing them from getting lost in the carpet, falling off the table and rolling away, or blending in with the other clutter on the workbench.

Whenever disassembling any firearms, special care is warranted where springs are concerned. Exposing, loosening, or removing compressed springs can sometimes be challenging because they have a tendency to go flying in unknown directions. It is usually best (and safest) to cup a hand or cloth rag over any exposed spring until its complete removal or the reassembly is concluded.

It is often recommended that shooters keep both eyes open while aiming and firing their weapons, as opposed to squinting or closing one eye. Not only does this allow a wider field of view, but it also makes target acquisition quicker and reduces eyestrain.

I learned in Army basic training 28 years ago that using the very tip of the trigger finger when shooting a rifle tends to result in better accuracy than wrapping the finger around the trigger, because it depresses the trigger more directly straight back as opposed to pulling it to one side or the other and in so doing pulling the shot off target.

Here is something important to consider whenever using any type of support or rest to aid in shooting: it is better to lay the *forearm* of a rifle on a rest (preferably padded) rather than setting the barrel directly on a car door, fence, table, or other fixture. Whenever a round discharges, there is always a degree of recoil and vibration generated, and the accuracy of the shot will likely be disturbed if the barrel is in direct contact with anything solid.

Shooters who shoot a lot and who like to keep their weapons clean will often save their old, worn-out shirts for gun rags. Gun rags can be kept dry and stored within the kit inside plastic zip-seal bags.

Large-bore rifles, muskets, and shotgun barrels tend to collect dust and cobwebs inside their bores over time while resting in a gun rack or being stored in a bedroom closet. With muzzleloaders, it is more difficult to see exactly what has accumulated down the barrel. I found that plugging the muzzle with a wine cork when not in use helps keep spiders and dust out, as well as the inside of the barrel dry in wet climates.

Finally, any readers who are new to firearms and the shooting sports can learn the essentials about how to avoid most hazards by taking a firearms safety course or two before heading out to the range with their new weapons. For courses in your area, check with your local gun range, gun shop, or police department, or check out the training page at the National Rifle Association's website. There you can select the course or courses that best suit your needs and find the location for those nearest your home.

www.ingramcontent.com/pod-product-compliance
Lightning Source LLC
Chambersburg PA
CBHW081408270326
41931CB00016B/3415